Mirela Aline Real de Lima

Cytogenetic analysis protocol for canine tumors

Mirela Aline Real de Lima

Cytogenetic analysis protocol for canine tumors

The step-by-step protocol for error-free cytogenetic analysis

ScienciaScripts

Cover image: www.ingimage.com

This book is a translation from the original published under ISBN 978-3-330-76904-5.

Publisher:
Sciencia Scripts
is a trademark of
Dodo Books Indian Ocean Ltd. and OmniScriptum S.R.L publishing group

120 High Road, East Finchley, London, N2 9ED, United Kingdom
Str. Armeneasca 28/1, office 1, Chisinau MD-2012, Republic of Moldova, Europe
Managing Directors: Ieva Konstantinova, Victoria Ursu
info@omniscriptum.com

Printed at: see last page
ISBN: 978-620-8-60425-7

SUMMARY

To the best family in the world:

My parents and siblings for their endless support, understanding, affection and love.

"Of all the animals we know, the dog is the one that has bonded with us the most, whether it's princes who give him plenty of food and a feather bed, or beggars who sleep in the open and can only offer him a small portion of their own crumbs, his affection and dedication are identical, and with equal love he licks his jeweled hand and his trembling fingers, consumed with sickness and hunger."

Théo Gygas, in "The Dog in Our House"

ACKNOWLEDGMENTS

First of all, I thank God for putting me on a good path and for helping me to walk it.

To Professor Maria Lùcia Zaidan Dagli, for believing in the girl who appeared with a desire and made a dream come true; also for trusting in the work of a biomedical doctor in the midst of Veterinary Medicine and for the way she treats all her "children".

I am infinitely grateful to my parents, Dolores and Manoel, for having united their gametes for the sake of genetics, not on purpose. To my siblings, Ygor and Marianne, for the moments of joy, unity and consideration. My family was very important to me in carrying out this work. Thank you so much for standing by me and believing in my ability. I love you all very much.

To Filipe, my boyfriend, friend and companion, and his family, for all the happy weekends, memorable fishing trips and dedication to me. Thank you for always being by my side.

Special thanks to my cousin Paulo, for his patience and didactic teaching of computer science and its magical shortcuts.

To the lab technicians, Marguiti and Cyntia, for all their scientific and technical support and, above all, for their friendship. Incredible, admirable and kind people; today, friends for the future.

To my colleagues at the Experimental Oncology Laboratory who have become friends. For all the help, teachings, consideration, friendship, concern and for all the moments of joy we were able to share. Bruno, Lucas, Tereza, Patricia, Màrcia, Priscila, Juliana, Natâlia, Ivone, Luise, Marcelo, Daniel, Heidge, Zé, Gabriela, Heloisa, Kâtia Pinello, Kâtia Kimura, Fàbia. Among the special people, there are the most special... Bru, Lucas and Tê, I mean you!

I would like to thank the staff of the Department of Pathology for all their help, friendship and the many moments of relaxation we were able to spend together.

To the staff and friends at HOVET, who often went above and beyond to help me carry out the experiments. Thais and Leda: there are no words to define the strength you have given me.

To all the FMVZ staff who were directly and indirectly involved in the completion of this master's degree.

To my angel friends at the Institute of Biosciences in the Department of Genetics and Evolutionary Biology, who taught me much of what I know about genetics.

To Professor Yatiyo, who took me in and allowed me to meet the angels!

I would like to say a huge thank you to Rodrigo, Renata and Màrcia, who I now consider great friends, for their company, teachings, patience, all the techniques, protocols and all the moments of laughter and coffee together. To my colleagues at Bio, who made my working moments happier.

To all the people who made my master's degree one of the happiest phases of my life; to all the people who cared about me and for all the things I learned during this phase.

Special thanks to the dogs who took part in this work and to their owners, who authorized the use of the materials.

I would like to thank FAPESP for their financial support for this dissertation.

SUMMARY

REAL-LIMA, M. A. **Compared cytogenetic analysis in mast cell tumors: a special focus on Boxer breed.** [Compared cytogenetic analysis in mast cell tumors: a special focus on Boxer breed. 2009. 115 f. Dissertation (Master of Science) - Faculty of Veterinary Medicine and Zootechny, University of Sao Paulo, Sao Paulo, 2009.

Many techniques are currently being used in the field of cytogenetics to identify or prevent potentially heritable genetic imbalances. Some of these techniques are only used to diagnose diseases in humans, but it would be important if they were also applied to domestic animals, with the aim of associating a genetic cause with the appearance of certain diseases, such as mastocytomas. Mastocytomas are neoplastic skin formations originating from mast cells and are common in dogs. Some breeds are more susceptible to mastocytomas, and the Boxer has the highest incidence in this species. The aim of this study was to comparatively evaluate the karyotype of Boxer dogs and look for a possible correlation with the development of mastocytoma. Cytogenetic techniques such as C, G and Ron banding and conventional Giemsa staining were used. Cultured cells from peripheral lymphocytes and Boxer mastocytomas were used for these analyses. The material was analyzed and photographed using a Zeiss® immersion microscope with a 100x objective and a green filter, equipped with Ikaros® cytogenetic analysis software. Numerical and structural changes were taken into account. A total of 828 mitotic metaphases from peripheral lymphocytes and mastocytoma cells were analyzed. The analysis of metaphases from peripheral lymphocyte cell cultures showed no structural alterations and the numerical analysis showed only one metaphase with 2n=77, out of 476 metaphases analyzed. In the analysis of the metaphases of tumor cells, among all the stains, it was observed that 70.9% showed 2n=78, 17.4% showed 2n=77; 9.8% 2n=76. Five metaphases (1.3%) showed 2n=79 and two cells showed 2n=75. All these numerical changes are related to structural alterations of the centric fusion type. The results show that due to uncontrolled growth, tumor cells promote rearrangements in an attempt to control cell division. Furthermore, the improvement of diagnostic techniques in cytogenetics promotes knowledge of the tumor biology of canine mastocytoma, and consequently enables the development of new therapeutic strategies.

Keywords: Mastocytoma. Boxer. Cytogenetics. Chromosomes. Alterations chromosomes.

1 INTRODUCTION

Man's relationship with small domestic animals, especially dogs, is ancient. Many dog breeds have developed from crossbreeding animals of different sizes, and these crossbreeds have resulted in breeds with a greater genetic predisposition to developing certain diseases.

The Boxer breed has German origins and was developed at the end of the 19th century by a group of German cynophiles who sought to obtain, by crossing an ancient type of dog known as the Bullenbeisser with the English Bulldog, a breed with homogeneous and elegant characteristics.

Boxer dogs are described as carriers of genetic "defects", a phrase justified by the high rate of development of diseases, especially neoplasms.

Some types of skin and soft tissue tumors are frequently found in domestic animals due to a series of intrinsic characteristics of this tissue. In addition, animals are subject to genetic factors and internal agents involved in tumor production (PULLEY; STANNARD, 1990) and the skin is exposed to a variety of chemical and physical factors (HUEPER, 1963). One known neoplasm is mastocytoma.

Mastocytomas are neoplastic cutaneous formations originating from mast cells with genetic alterations (SELYE 1965; VAIL 1996; SOTER 2000) and make up approximately 7 to 20% of cutaneous neoplasms in dogs (O'KEEFE, 2003). These formations develop in many places, but are most commonly found in the dermis, subcutaneous tissue and mucous membranes (MISDORP, 2004).

Once the mastocytoma has been diagnosed, the animal usually undergoes surgery to excise the tumor mass and then undergoes the available treatments, such as chemotherapy and radiotherapy, in order to establish a good quality of life.

There are now advanced techniques, previously only used for human treatments, which are being used as tools for diagnosing and treating some animal diseases.

One of these techniques is cytogenetics, which has been used to elucidate previously unsolved enigmas. Using a small number of cells, cytogenetics makes it possible to carry out a complementary analysis to the histopathological analysis, for example in the case of mastocytomas.

Cytogenetic analysis makes it possible to find out about chromosomal alterations, so it is possible to diagnose a disease and prevent it from being passed on to future generations by crossing affected animals, which prevents the spread of these alterations.

Published works highlight advances in the discovery of chromosomal alterations involved in canine neoplasms with the use of cytogenetic techniques, which make it possible to better analyze more specific portions of the chromosomes (WINKLER et al., 2005).

The aim of this study was to analyze and describe some chromosomal alterations found in cells from mastocytoma tumor masses, which may be associated with their development.

The following pages will describe tumor behavior and cytogenetic methods, as well as an overview of the literature and historical aspects of what is already known about canine genetics and cytogenetics, to serve as a technical and scientific basis for the reader to understand the findings of this dissertation.

2 LITERATURE REVIEW

The following pages will present an overview of the literature and historical aspects on the subjects involved in this work, so that the reader can familiarize themselves with the topic.

2.1 MASTOCYTOMAS

Mastocytomas are neoplastic cutaneous formations originating from mast cells with genetic alterations (SELYE 1965, VAIL 1996, SOTER 2000) and make up approximately 7 to 20% of cutaneous neoplasms in dogs (O'KEEFE, 2003). These formations develop in many places, but are most commonly found in the dermis, subcutaneous tissue and mucous membranes (MISDORP, 2004).

Mastocytomas have an unknown origin, but there is a correlation between racial predisposition, chronic inflammation, viral action in the development of the disease (GRAHAM and O'KEEFE, 1998) and point mutations in the "c-kit" gene in neoplastic dog mast cells (REGUERA et al., 2000). The main mutation found was a duplication that causes constitutive phosphorylation of the receptor, without the need for binding to Stem Cell Factor, "SCF". This explains the uncontrolled growth of tumors and the positive relationship between duplications and mastocytoma malignancy (ZEMKE et al., 2002).

There are reports that sites with foci of chronic inflammation give rise to mastocytomas, as well as speculation that viruses may play a role in the development of mastocytomas (VAIL, 1996; GRAHAN and O'KEEFE, 1994). This hypothesis is based on experimental studies in which dogs developed mastocytomas after receiving injections of cell-free tumor extracts (O'KEEFE, 2003).

Mast cells are a normal component of the immune system and are important in the inflammatory response to tissue trauma. The cytoplasmic granules found in mast cells contain biologically active substances such as heparin, histamine, platelet activating factor and eosinophilic chemotactic factor (JANEWAY et al., 2002; METCALFE et al., 1997). Mast cells have a hematopoietic origin, migrate to connective tissue when they are mature and retain the ability to proliferate after maturation (ZAPPULLA et al., 2002): normal canine mast cells resemble human mast cells (LAZARUS, 1986).

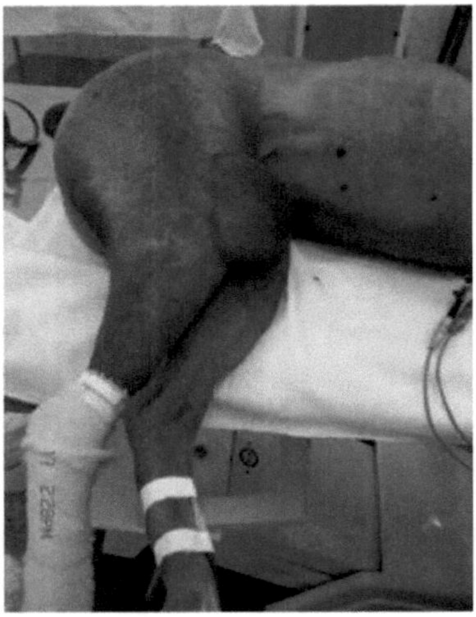

Figure 1 - Seven-year-old male Boxer with grade II mastocytoma on the right hind limb. The tumor was surgically excised at the Surgery Department of the FMVZ-USP Veterinary Hospital. The animal had a 12-month interval for tumor recurrence

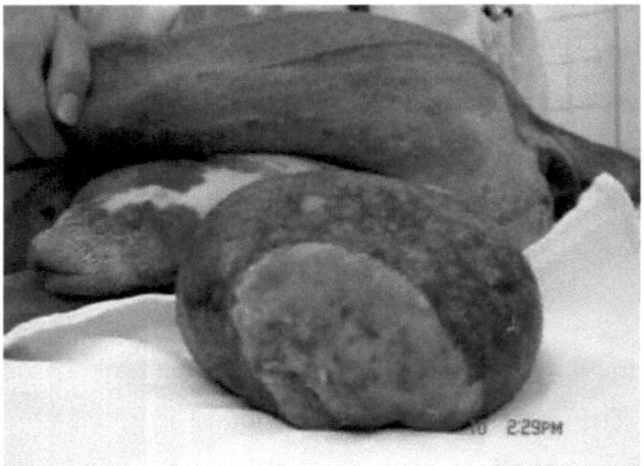

Figure 2 - Nine-year-old male Boxer with grade III metastatic mastocytoma in the scrotum, bilateral inguinal lymph nodes and left iliac lymph node. Surgical excision of the scrotum and inguinal lymph nodes was performed at the Surgery Department of the FMVZ-USP Veterinary Hospital. The animal underwent post-operative chemotherapy and survived four months after surgery

2.1.1 PREDISPOSITION

Racial predisposition mostly involves dogs of no defined breed, but it also affects the Boxer, Boston Terrier, English Bulldog, English Bull Terrier, Beagle, Labrador and Schanauzer breeds (VAIL 1996, O'KEEFE 2003). It occurs mainly in dogs with an average age of 8-9 years (VAIL, 1996).

In general, Boxer dogs are highly susceptible to developing neoplasms, and given that mastocytomas are the most common neoplasms in dogs, this disease affects Boxer dogs more specifically. In these animals, mastocytomas are generally well-differentiated and have a more favorable prognosis (O'KEEFE, 2003).

1.1.2 DIAGNOSIS

Fine needle aspiration cytology is efficient for establishing the initial diagnosis of mastocytoma, but the definitive diagnosis, directly related to prognosis, is the histopathological examination, which is fundamental for classifying the grade of the tumor (MACY, 1986).

Mastocytomas are classified histomorphologically into three categories (grade I, II and III), according to the degree of differentiation of the mast cells present. This classification is the most widely used and was developed by PATNAIK et al. (1984).

The degree of differentiation of tumors is related to the quantity and type of granules present in the tissue, with well-differentiated tumors containing more heparin than undifferentiated tumors, which have a higher histamine content (LEMARIÉ et al., 1995). With the release of histamine and stimulation of the stomach's H2 receptors, there is increased secretion of hydrochloric acid and increased gastric motility, causing duodenal ulcers in 80% of animals with mastocytoma (MADEWELL and THEILEN, 1987). The percentages observed were 36% of mastocytomas classified as grade I, 43% grade II and 20% grade III (PATNAIK et al., 1984).

2.1.3 HISTOPATHOLOGICAL CLASSIFICATION OF MASTOCYTOMA

Grade I or well-differentiated mastocytoma (Figure 3): the tumor is usually small, confined to the dermis and is not encapsulated. The surface of the neoplasm is often not ulcerated. The neoplastic cells resemble normal mast cells and are arranged in cords and/or thin layers. The cytoplasm is discrete with prominent granules that are stained blue-grey after hematoxylin-eosin staining. Mitoses are rare or absent. Eosinophils are found in significant numbers. Some tumors show small foci of collagenolysis around degenerated eosinophils (flame-like figures); in this condition, there is often little stroma in the collagen bundles, with a variety of edema; the nucleus is round or oval; it is small, with an almost imperceptible nucleolus. Mitoses are infrequent or absent. Perivascular hyalinization is common. Tumor necrosis is rare, but infarcted areas may appear (GOLDSCHIMIDT and SHOFER, 1992; GROSS, IHRHE and WALDER, 1992). Grade I mastocytoma represents the majority of tumors in dogs and usually presents benign behavior (GROSS, IHRHE and WALDER, 1992;

SCOTT, MILLER and GRIFFIN, 1996).

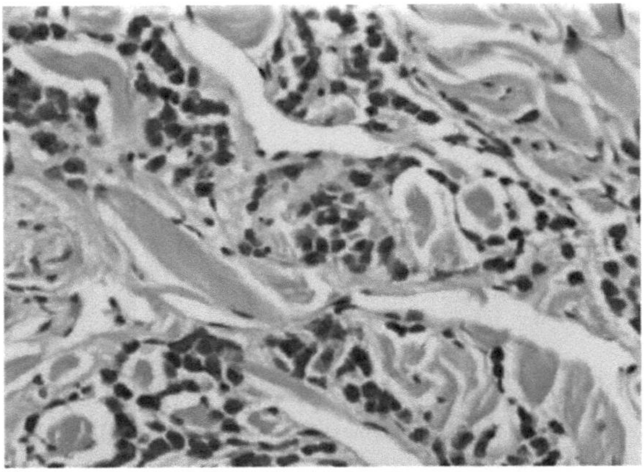

Figure 3: Histology of grade I mastocytoma (200x, HE stain).

Grade II or moderately differentiated mastocytoma (Figure 4): the tumor is larger and less circumscribed than the previous one. It usually affects the dermis and subcutaneous tissue. The neoplastic cells are still arranged in cords and sheets, although more disorganized and with an infiltrative growth pattern; they are larger, may be pleomorphic and have fewer cytoplasmic granules. Areas of necrosis and ulceration of the tumor surface are commonly found. The nucleus is larger than that of a normal mast cell and has pleomorphic characteristics; the nucleolus is small, although it can be identified. Mitotic activity is considered low to medium. The amount of stroma and eosinophils decreases. Flame-like figures are rarely observed (GOLDSCHIMIDT and SHOFER, 1992; GROSS, IHRHE and WALDER, 1992). Grade II mastocytomas account for around 20% of mastocytomas in dogs (GROSS, IHRHE and WALDER, 1992). The diagnosis of grade I and II mastocytic tumors is usually clear and straightforward (GROSS, IHRHE and WALDER, 1992). The prognosis is also more favorable than that of Grade III, which is described below:

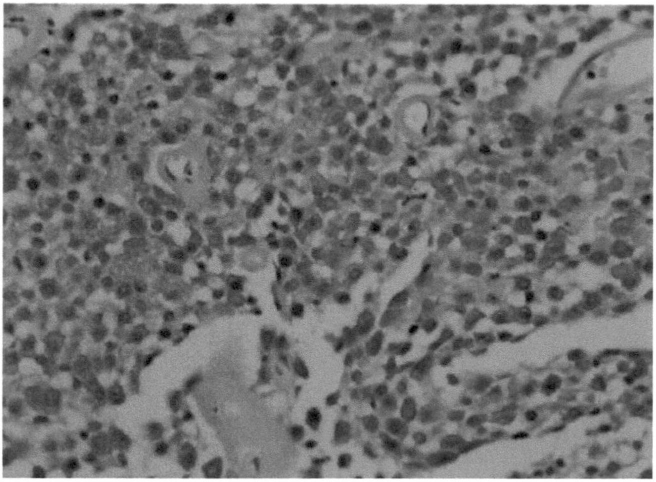

Figure 4: Histology of grade II mastocytoma (200x, HE stain).

Grade III or poorly differentiated mastocytoma (Figure 5): Usually, the tumors are large and ulcerated and extend beyond the subcutaneous tissue. The neoplastic cells are organized in solid layers and the granules are not discernible without special stains such as Giemsa or toluidine blue; they are round, polygonal and quite pleomorphic. The cell margin is not well defined. The nucleus is large, round to oval and vesicular; the nucleolus is often prominent. Mitotic activity varies from medium to high and several atypical mitoses are observed. There are few eosinophils and the stroma is scarce or non-existent (GOLDSCHIMIDT and SHOFER, 1992; GROSS, IHRHE and WALDER, 1992). This type of tumor is less common and the mortality rate due to systemic complications is high (GROSS, IHRHE and WALDER, 1992). Expression of the c-kit oncogene is higher in Grade III or undifferentiated tumors (REGUERA et al.; ZEMKE, YAMINI and YUZBASIYAN-GURKAN, 2000).

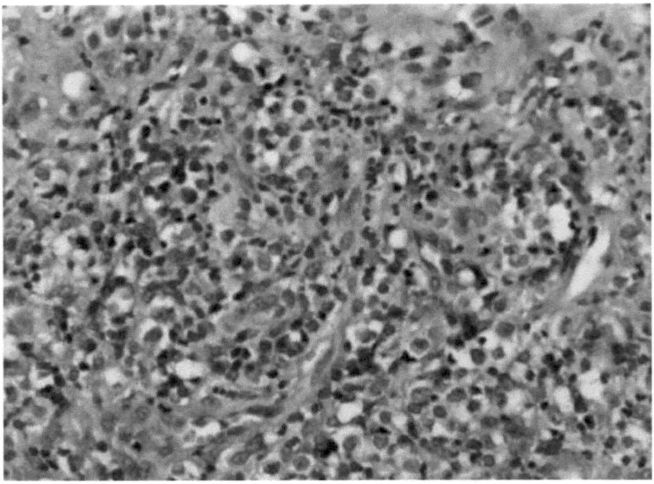

Figure 5: Histology of grade III mastocytoma (200x, HE stain).

1.1.4 CLiNICAL SIGNS OF MASTOCITOMA

The clinical signs of mastocytomas depend on the location and size of the tumor, as well as the secondary systemic complications caused by the mastocytoma. Clinically, solitary and ulcerated nodules are most often seen on the trunk and limbs of animals (MEDWELL AND THEILEN, 1987).

On the other hand, part of the mastocytomas are seen as solitary masses on the skin of the trunk and perineal area, with an incidence of around 50% of cases, masses on the extremities, such as the muzzle and ears, in 40% and on the head and neck, affecting around 10% of animals (O'KEEFE, 2003).

Dermal mastocytomas are usually prominent, well-defined masses, which can become hairless over the region, ulcerated and erythematous, and subcutaneous mastocytomas can resemble lipomas (Graham and O'KEEFE, 1998).

There are reports of duodenal ulcers in up to 80% of dogs with mastocytomas and it is believed that this is related to the release of histamine, leading to possible signs such as anorexia, vomiting, diarrhea and melena (OGILVIE and MICHENER, 2000).

The regional lymph nodes, spleen and liver are the most common sites of metastasis (MEDWELL and THEILEN, 1987). Hά reports of metastases observed in the bone marrow, which suggests systemic involvement of the tumor (Lemarié et al., 1995), with pulmonary metastases being less frequent (GRAHAM and O'KEEFE, 1998).

Low-grade mastocytomas are less likely to recur or metastasize (O'BRIEN and WHITHRON, 1997), and dogs with these tumors have longer survival periods (O'KEEFE, 2003).

The biological behavior of mastocytomas depends on the staging and histological differentiation of the tumors, age and species affected, and as it is often difficult to classify the staging of tumors, all mastocytomas should be considered potentially malignant, with an incidence of approximately 50% of mastocytomas classified in this way (GRAHAM and O'KEEFE, 1998).

Progression time and speed of growth are important factors that must be taken into account to establish staging and subsequent treatment and prognosis, so early diagnosis of slow-growing tumors is associated with a better prognosis, even with simpler therapies (SIMÔES, 1994).

The location and speed of tumour growth can help predict biological behaviour, characterizing tumours in the preputial, inguinal and perineal areas as having a more aggressive tendency (O'KEEFE, 2003). On the other hand, mastocytomas located in a specific area, such as the intestinal mucosa, and which grow slowly, may have a better prognosis (MEDWELL and THEILEN, 1987).

1.1.5 TREATMENT OF MASTOCYTOMAS

Once the tumor has been classified, mastocytoma treatment may include surgery, radiotherapy, chemotherapy or some combination of the three. The type of treatment instituted depends primarily on the histological grade and clinical stage of the tumor, which allows surgery to be combined with radiotherapy and/or chemotherapy (GRAHAN and O'KEEFE, 1994).

With skin manipulation, there may be degranulation of the mastocytoma, resulting in Darier's sign, which refers to a localized area of dermal edema and erythema (MURPHY et al., 2005). The release of heparin and proteolytic enzymes by mastocytomas during surgery can prolong clotting time and delay wound healing (JANEWAY, 2002).

Although they appear to be a discrete mass, mastocytomas generally infiltrate deep into the surrounding tissues, making comprehensive surgical excision mandatory (O'BRIEN and WHITHRON, 1997). Supportive therapy before, during and after surgery is essential (KITCHELL,

1993), so the chances of recurrence are reduced and the animal is not subjected to treatment again.

2.2 THE BOXER BREED

The Boxer breed has German origins and was developed at the end of the 19th century by a group of German cynophiles who sought to obtain, by crossing an ancient type of dog known as the Bullenbeisser with the English Bulldog, a breed with homogeneous and elegant characteristics.

Around 1890, a dog with the characteristics of today's Boxer showed up in Munich, a German city in the Bavarian region, and mated with an old bulldog. The result was a dog with a developed chin, high-set ears and strong bones.

As a result, these specimens were crossed in order to eliminate the excess white color from the bulldog, thus obtaining the desired blond or golden color. In addition, the main aim was to reduce robustness and heavy conformation, seeking a more agile and elegant conformation.

After more than a hundred years of perfecting the breed, what we see today is a harmonious, elegant and powerful dog. Compact and square in figure, its bones are heavy and its coat short. Its plastic musculature is powerfully developed and clearly defined, giving it an athletic and noble appearance.

Of medium to large stature, males should measure around 60 cm, with males over 60 cm weighing more than 30 kilos. The slightly smaller females (Figure 6) measure around 55 cm and those measuring over 56 cm should weigh more than 25 kilos.

Boxer dogs are described as carriers of genetic problems due to their high rate of developing diseases, especially neoplasms.

Figure 6: Female Boxer dog.

2.3 GENETICS AND CANCER

The transformation of a normal cell into a neoplastic cell is called carcinogenesis, characterized as a process that involves at least three stages: initiation, promotion and progression (FABER, 1984).

The initiation phase is characterized by irreversible DNA damage which can be caused by chemical, physical (ultraviolet and ionizing radiation), biological (viruses) or genetic (heredity) agents (COTRAN; KUMAR; ROBBINS, 1989).

In the promotion phase, expansion is observed through the growth of phenotypically altered cells due to a stimulus to cell proliferation, which may be due to the release of a growth factor (FABER; SARMA, 1987).

The progression phase is characterized by the irreversibility that the genetic alterations confer on the cells, such as high proliferative potential, invasive capacity, metastatic capacity and biochemical alterations (PITOT; DRAGAN, 1991).

Cancer is not the result of a single event inside the cell. A series of events must take place for cancer to develop. A single cell can undergo an initiation event, but the progeny of this cell must be subjected to new genetic alterations for malignant cells to develop (SOLOMON et al., 1991).

In many cases, these genetic changes are tiny mutations in individual genes. In other cases, the genetic alterations are the result of large-scale changes in the structure of the chromosomes themselves (ARMITAGE et al., 1954).

There are three categories of genes involved in carcinogenesis: proto-oncogenes (growth promoters), anti-oncogenes (tumor suppressors) and apoptosis regulating genes (COTRAN; KUMAR; ROBBINS, 1989).

Proto-oncogenes encode proteins that require external stimuli and are subject to regulators. When proto-oncogenes are altered in such a way as to give rise to neoplastic cells, they are called oncogenes.

Oncogenes encode oncoproteins that resemble the products of proto-oncogenes, but are devoid of regulatory elements and their production in transformed cells does not depend on growth factors or other external signals. Oncogenes can be present at all stages of cell proliferation (FABER; SARMA, 1987).

Apoptosis-regulating genes regulate cell death by controlling oxidative pathways. By perpetuating cell life, the overexpression of apoptosis-regulating genes promotes the emergence of mutations that affect growth-promoting genes and anti-oncogenes (COTRAN; KUMAR; ROBBINS, 1989).

2.3.1 CHANGES

Mutations do not occur as rarely as they might seem. Their frequency is relatively high, but as they are generally deleterious, DNA repair mechanisms prevent most mutations from becoming established (SOLOMON et al., 1991).

Mutations are produced, among other things, by mutagenic agents, which interfere with DNA self-duplication or mRNA transcription, causing errors in nucleotide sequences. Mutagens are factors that can increase the frequency of mutations. Chemical substances such as "mustard gas" and nitrous acid (HNO_2), aerosols, food colorings and some components of cigarette smoke can also increase the frequency of mutations, altering a cell's genetic heritage and potentially leading to the development of different forms of cancer.

Base tautomerism is incorrect pairing in the DNA arrangement. Instead of pairing adenine-thymine (A-T) and guanine-cytosine (G-C), there can be adenine-cytosine (A-C) and thymine-guanine (T-G).

Ionizing radiation: in most studies, the frequency of point mutations is directly proportional to the dose of irradiation (X-rays, gamma rays and cosmic rays).

Ultraviolet radiation is absorbed by purines and pyrimidines, forming pyrimidine hydrates and pyrimidine dimers (thymine dimerization - where two nearby thymines from the same DNA strand bind, impairing the fit of the double strand).

Acridines are compounds that intercalate between the double strand, causing the insertion or deletion of one or more base pairs. There is an alteration in the DNA reading module.

Alkyls transfer methyl or ethyl groups to the reactive sites of the bases and phosphates in the DNA chain. A well-known and well-studied example of mutation involves the hemoglobin molecule, which is a complex protein made up of four globin protein chains combined with the complementary heme iron.

All living beings are subjected to several of these agents on a daily basis. The relative stability of genetic material is due to the existence of a group of repair enzymes, which permanently "patrol" DNA molecules looking for alterations in the sequence of its nucleotides. Most of the time, these alterations are detected and repaired. There are various repair mechanisms, such as nucleotide excision, where an enzyme cleaves and removes the region where the mutation occurred. The resulting void is then filled correctly by the action of the DNA-polymerase enzyme, which uses the DNA chain as a template for repair.

2.4 CYTOGENETICS

Cytogenetics encompasses any study of the chromosome, whether isolated or as a whole, condensed or distended, in terms of its morphology, organization, function and replication, as well as its variation and evolution (GUERRA et al., 1988).

Karyotype analysis or cytogenetic studies can be carried out on any cell tissue, as long as it is collected properly. The tissues commonly used, either because of their ease of access or their specificity, are peripheral blood, bone marrow, skin, solid tumors and fetal tissues.

2.4.1 HISTORICAL ASPECTS OF CYTOGENETICS

The first ideas about chromosomes emerged at the end of the 19th century, when the first studies on mitosis were carried out. In 1866, Haeckel proposed that the cell nucleus was the main agent responsible for cell division. However, the first scientist to describe the process of mitotic cell division clearly was the German zoologist Anton Schneider, in 1873.

Subsequently, other researchers expanded their studies on mitosis. Important contributions were made by Strasburger in 1875 and by Flemming from 1879-1882, the latter who coined the term "mitosis" (process of cell division, from the Greek mitos = filament). From then on, mitotic division was better understood in animals and plants.

Although studies of chromosomes had become frequent, the idea that they were involved in genetic inheritance only emerged in 1887, in the work of Weismann, who introduced them into the theory of inheritance. This included the following points:

1. "nuclear substance" controls the shape and function of each cell, and divides (mitosis) generating products equal to itself.

2. *Eggs* need to lose half of their "nuclear substance" in the polar body before fertilization, and need to be completed with exactly the same amount of "nuclear substance" from the sperm.

3. Because sexual reproduction depends on the nuclei of the egg and the sperm being joined in each generation, the "nuclear substance" of the germ cells of both must have been halved. (This proposition was formulated after the process of meiosis was understood).

4. There are no substantial differences in the "nuclear substance" of egg and sperm cells.

5. Sexual reproduction can be understood as producing variability between individuals, where natural selection can act.

Weismann's theory considers all these points (which can be observed in nature), even though it was formulated in complete ignorance of Mendel's laws. These were only rediscovered in 1900 by three scientists: Correns, de Vries and Tschermak, in a series of cytology studies. With the revival of Mendel's laws came the theory of chromosomal inheritance, which relates chromosomes to the Mendelian laws of inheritance.

This was a milestone for the study of chromosomes and, at the beginning of the 20th century, cytology and genetics began to integrate their knowledge into a common area, later called Cytogenetics.

2.4.2 CHROMOSOMES AND DNA

Chromosomes (Kroma=color, soma=body) are spiral strands of chromatin, found in the nucleus of cells, made up of a long sequence of DNA and proteins. DNA contains several genes and other nucleotide sequences with specific functions in the cells of living beings. Each chromosome is a DNA molecule.

The first researcher to observe chromosomes was Karl Wilhelm Von Nageli in 1842, and their behavior was described in detail by Walther Flemming in 1882. In 1910, Thomas Hunt Morgan proved that chromosomes are the carriers of genes.

For karyotypic analysis of mammals, mitotic metaphase chromosomes are the most suitable, mainly due to their high degree of condensation, which facilitates counting the diploid number, defining morphology and identifying chromosomes (GUERRA and SOUZA, 2002).

Mitosis is the process of cell division which distributes the chromosomes and cytoplasmic components of the mother cell equally between two daughter cells. This process is responsible for the multiplication of unicellular organisms and the growth of multicellular organisms by increasing the number of cells.

Mitosis is divided into four stages: prophase, metaphase, anaphase and telophase. It is precisely in the metaphase stage that the chromosomes are best observed due to their greater concentration and individualization. Anti-mitotic substances are used to stop mitosis in metaphase. These substances bind to the proteins that form the achromatic spindle fibres (tubulins), preventing their polymerization and consequently suppressing the formation of the fibres, or even inactivating the spindles already formed. The most commonly used antimitotic substance is colchicine.

The chemical composition of chromatin includes proteins, DNA and RNA (the latter in smaller quantities). The proteins are of two classes: the basic proteins, called histones; and the acidic proteins, called non-histone chromosomal proteins.

Histones play an important role in structuring the chromosome. So much so that their proportion in the nucleus is 1:1 (weight/weight) in relation to the DNA itself.

The chromatin structure of eukaryotes is made up of repeated subunits called nucleosomes. These consist of a segment of DNA 146 nucleotide pairs long, wrapped around the somewhat cylindrical surface of the histone octamer, producing a roughly ellipsoid structure.

At least five levels of condensation are required to package the DNA(a) in the eukaryotic chromosome into a metaphase structure of a few micros in length, as shown below in figure 7.

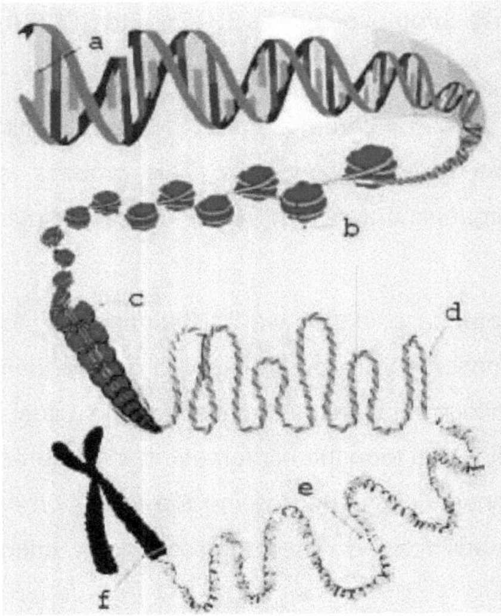

Figure 7: Illustration of DNA condensation levels.

1. The first involves the packaging of DNA as a spiral into nucleosomes approximately 10nm in diameter. This step involves an octamer of histones (b).

2. The second involves the solenoid structure, a second level of epiralling, producing a 30nm fiber (c).

3. *The* third are the "loops" of solenoids, attached to a central protein skeleton. This structure is approximately 300nm in diameter (d).

4. The fourth are "loops" of the protein skeleton, forming a giant, super coiled structure measuring 700nm (e).

5. Finally, at its maximum condensation, the chromosome is about 1400nm in diameter (f).

2.4.3 CHROMOSOME STRUCTURE AND TYPES

The typical metaphase chromosome is made up of two sister chromatids, one of which comes from the chromatin duplication process. The chromatids are attached by a thin region called the primary constriction or centromere.

The centromere divides the chromatid into two chromosome arms. Or it can be located in the terminal region of one arm, forming a chromosome with only one arm.

A secondary constriction can also be seen on some chromosomes, another differentiated condensation region. The segment cut off by the secondary constriction and anterior to the telomere (end of the chromosome arms) is known as a satellite.

Most chromosomes are monocentric, i.e. they only have one centromere. In nature, you can find dicentric, tricentric, acentric, etc. chromosomes. The structure of a chromosome is shown in figure 8.

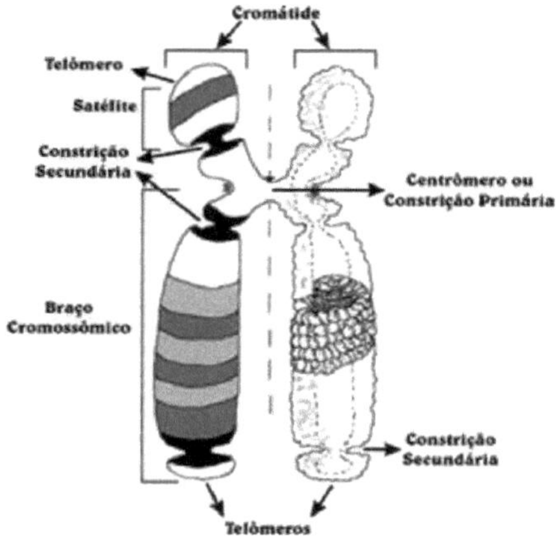

Figure 8: Schematic illustration of the structure of a chromosome.

Classifications of chromosome types (Figure 9) are based on the location of the centromeres. One example is the classification by Levan et al. (1964), which considers four chromosome types in relation to the arm ratio. The ratio is calculated by dividing the measurement of the largest arm (q) by that of the smallest arm (p).

Chromosome arm ratio:

Metacentric RB = 1.00 - 1.70

Submetacentric RB = 1.71 - 3.00

Subtelocentric RB = 3.01 - 7.00

Acrocentric RB > 7.01

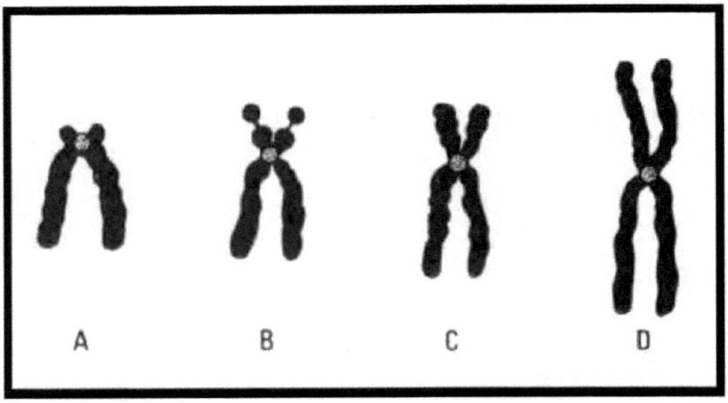

Figure 9: Schematic illustration of the types of chromosomes according to the classification of Levan et al. (1964): (A) acrocentric chromosome; (B) subtelocentric chromosome; (C) submetacentric chromosome; (D) metacentric chromosome.

Secondary constrictions are also particular regions of chromosomes. These are usually the nucleolus organizing regions (RONs). The genes responsible for producing rRNAs, which form part of the nucleolus, are located in the RONs. Their importance lies in producing and processing the RNAs needed to synthesize all the cell's proteins. Ribosomal genes are present in multiple copies in the genome, i.e. it is a type of repetitive DNA.

2.4.4 THE CHARIOTYPE

Based on the shape and number of chromosomes in a species, its karyotype is established. The karyotype is the representation of the set of chromosomes present in a cell, arranged in pairs of homologues. The word karyotype is derived from the Greek words "karyon" (node) and "typos" (shape).

The representation of the karyotype can be a karyogram (image of the chromosomes) or an idiogram (diagram of the chromosomes), and provides substantial information for establishing relationships between species with regard to the organization of the chromosomes.

2.4.5 COLORING AND BANDING

In addition to the so-called conventional stains (Giemsa, Acetic Orcein, Schiff's reagent, hematoxylin/eosin, etc.), methodologies that identify "bands" can be applied to chromosomes.

A band is defined as a part of the chromosome in which it is clearly distinct from the adjacent segment, appearing lighter or darker. Chromosomes are visualized as a continuous series of light and dark bands.

Banding (C, G, Q, R, Ag-RON) is important for identifying homologous and homologous chromosomes, characterizing polymorphisms and even identifying kinship relationships between close species, distinguishing possible chromosomal rearrangements.

G bands are the most commonly used. This nomenclature comes from the Giemsa dye used, but it can be produced with other dyes. In G bands, the dark regions tend to be heterochromatin (contain many repeats), are late replicating and rich in adenine and thymine. The light regions tend to be euchromatin (transcriptionally active regions), with early replication and rich in guanine and cytosine.

The C-banding method allows the visualization of regions rich in constitutive heterochromatin, such as the centromeres of some chromosomes, highlighting those rich in guanine and cytosine.

Through the silver staining protocol, it is possible to observe the Nucleolus Organizing Regions (Rons), where the loci of the genes for ribosomal RNA (rRNA) are located. Rons only show an affinity for silver when they are transcriptionally active.

2.4.6 CHROMOSOMAL CHANGES

Each species has a specific number of chromosomes. Due to the occurrence of irregularities during nuclear division, cells or entire organisms can have aberrant genomes. Such chromosomal alterations can include entire genomes, entire isolated chromosomes or only parts of chromosomes. Thus, chromosomal alterations can be numerical or structural, and involve one or more autosomes, sex chromosomes or both. Every alteration has an important effect on development, as changes to the normal nucleus can result in phenotypic alterations.

2.4.6.1 NUMERICAL CHANGES

Numerical chromosomal aberrations (heteroploidies) include cases in which there is an increase or decrease in the number of chromosomes in the usual body. Variations that affect entire sets of chromosomes are defined as euploidies and those that result in the loss or addition of entire isolated chromosomes are aneuploidies.

Euploidies are numerical chromosomal alterations characterized by the presence of an extra haploid set in addition to the basic diploid number of chromosomes. The alterations can generate individuals with haploid (n), diploid (2n), triploid (3n), tetraploid (4n) or polyploid (when there are several excess genomes) characteristics.

Aneuploidy refers to the situation in which the number of chromosomes is not an exact multiple of the haploid number characteristic of the species. This anomaly is due to the non-separation (or non-segregation) of one or more chromosomes into the daughter cells during the cell division process. Non-segregation in mitosis is the result of the failure of the centromere to break at the beginning of anaphase or the loss of a chromosome because it has not attached to the spindle. When, as a result of this process of non-segregation, one chromosome of a given pair is missing, i.e. when the number of chromosomes in the cell is 2n - 1, the cell is said to have monosomy for this chromosome. If two elements of the same pair are missing (2n - 2), there is nullisomy. If, on the other hand, there is an increase in the number of chromosomes in a given pair, the cell will be polysomic for the chromosome in question; it will be trisomic, tetrasomic, pentasomic, etc., depending on whether it has 1, 2 or 3 extra chromosomes, in which case its chromosome number will be called (2n + 1), (2n + 2), (2n + 3), etc.

2.4.6.2 STRUCTURAL CHANGES

Structural chromosomal alterations include cases in which one or more chromosomes show alterations that affect the molecular structure of the DNA.

Generally, structural changes require breaks in the chromosomes, which can occur at various points in a chromosome or in the whole set. A broken end can join up with any other broken end, resulting in new arrangements between these chromosomes.

There are six basic types of processes that determine structural changes in an individual's genome: deletion, duplication, inversion, translocation, transposition and centric fissure/fusion.

Deletions occur due to the loss of chromosome fragments, which can exclude one or many genes and can originate from simple chromosome breakage or elimination (Figure 10).

Duplication occurs when a chromosome segment appears more than twice in a normal diploid cell. The segment can be attached to a chromosome or as a separate fragment. Through the segregation of these chromosomes in the gametes, duplications can be passed on to subsequent generations (Figure 11).

Inversion occurs when part of the chromosome detaches (due to chromosome breakage), rotates 180° and is re-inserted so that the genes are in reverse order. Some inversions result from the tangling of filaments during meiotic prophase. Inversions can include the centromere (pericentric inversions) or not include the centromere (paracentric inversions) (Figure 12).

Translocation occurs when part of a chromosome breaks off and attaches itself to a part of a non-homologous chromosome. This type of alteration can be of two types: 1) Simple, when the broken fragment of one chromosome is inserted into another non-homologous chromosome, or 2) Reciprocal, when the segments are exchanged between two non-homologous chromosomes (Figure 13).

Transposition refers to the exchange of segments from one place on the chromosome to another. Generally there are no major changes, as the genomic content is not altered, but pairing can be difficult during cell division (Figure 14).

Centric fusion and fissure are changes where the chromosome number can increase or decrease without any variation in the amount of DNA. This occurs due to the chromosome breaking at the centromere, giving rise to two acrocentric chromosomes (fission), or the reverse process (fusion) (Figure 15).

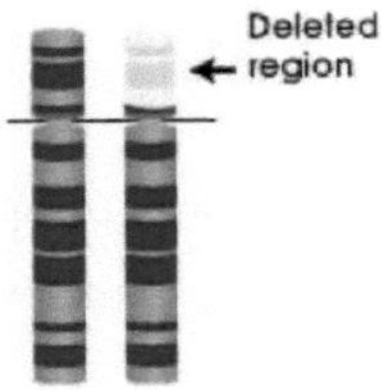

Figure 10 - Illustration of the deletion process

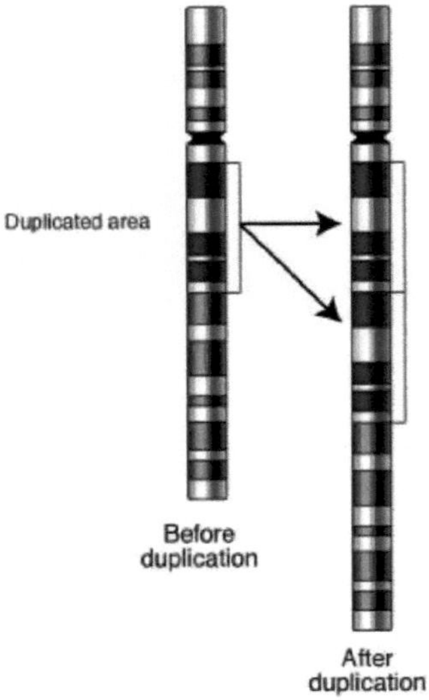

Figure 11 - Illustration of the duplication process

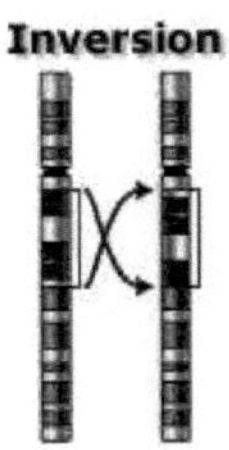

Figure 12 - Illustration of an inversion process

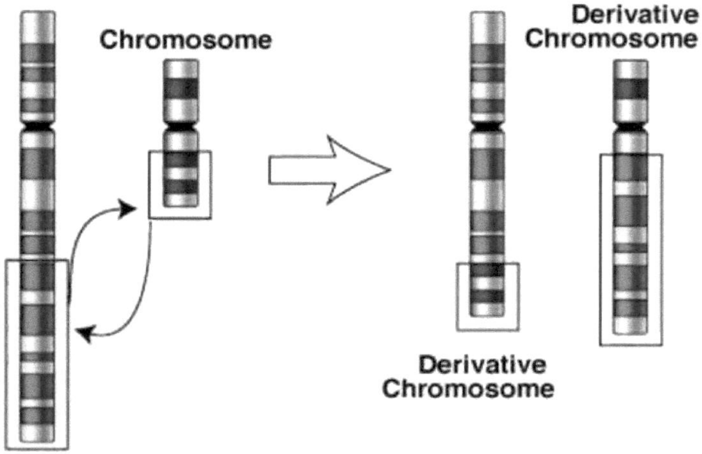

Figure 13 - Illustration of the translocation process

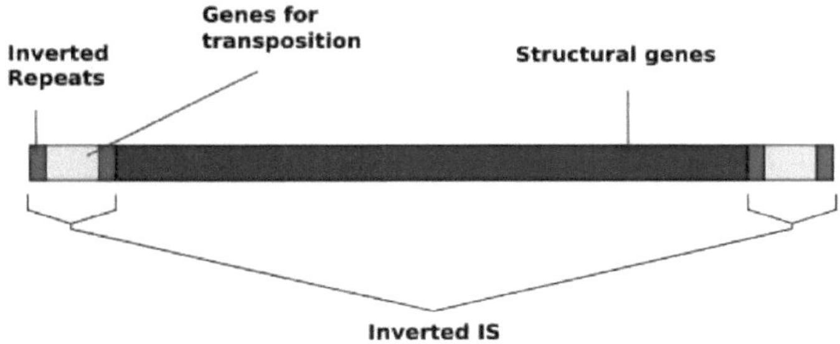

Figure 14 - Illustration of the deletion transposition process

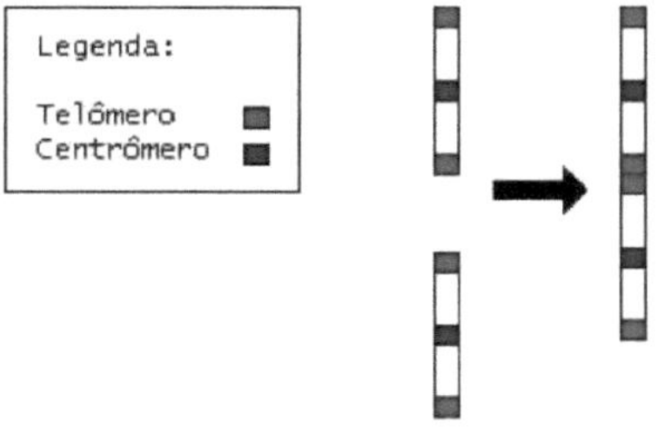

Figure 15 - Illustration of the fusion process

2.4.6.3 OTHER CYTOGENETIC CHANGES

Most of the DNA in cells is located in the chromosomes, but a small amount is found in extrachromosomal elements (EE), such as mitochondrial DNA and small covalently closed circular molecules of different types, including double minutes (dmin).

Double minutes (dmin) (Figure 16) and homogeneously stained regions (HSRs) (Figure 17), which occur in many tumor types, are gene amplifications and correspond to an inappropriate concentration of oncogenes. Double minutes are extrachromosomal structures of circular DNA (GAUBATZ, 1990), consisting of 1 to 2 million base pairs that replicate autonomously, approximately once per cell cycle, and segregate randomly to daughter cells, due to the absence of centromeres (CARROLL et al., 1987, 1993).

HSRs are amplified intra-chromosomal structures that are not necessarily in the native locus of the gene. Both contain genes that give a selective proliferation advantage to the tumor cells where they occur. It is hypothesized that DNA regions are excised from the chromosome as a result of recombination events at the replication fork (MITELMAN et al., 2007). Depending on their size, these amplified structures may contain some chromosomal genes and would result in a deletion of these genes from the corresponding loci. This model also predicts that episomes multimerize to form larger circular structures, the dmin, which can be detected cytologically. Subsequent to their formation, these extrachromosomal DNA elements can integrate into the chromosome, resulting in HSR.

Biological selection for the generation and maintenance of amplified DNA sequences in tumor cells is believed to be driven by increased copy number and increased expression of target genes or genes within a larger region of amplified DNA.

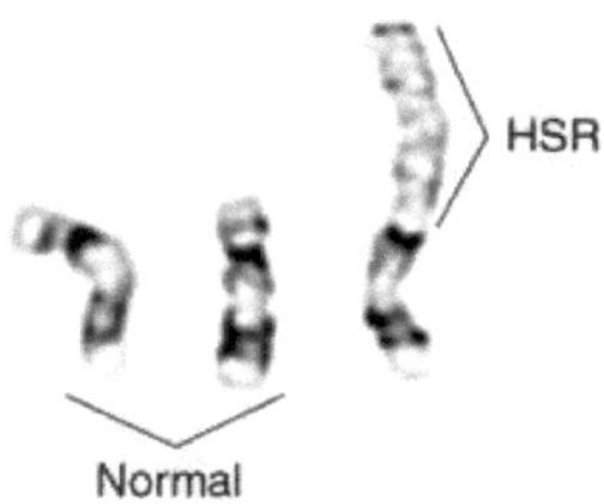

Figure 16 - HSR visualization on chromosome

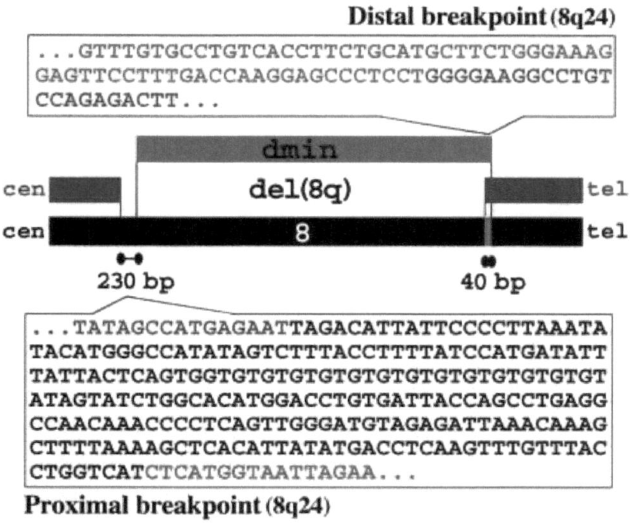

Figure 17 - Schematic representation of Double minutes

2.4.7 CANINE CYTOGENETICS

Over the last 100 years, several reports on the description of the karyotypic pattern of the dog have been published. In many reports, the diploid number ranged from 50 to 78 chromosomes (AHMED 1941; MINOUCHI 1927, PAINTER 1925; RATH 1894). Due to the development of new techniques, such as the hypotonization of cells in culture (HSU and POMERAT, 1953), the diploid number of 78 chromosomes, of which 76 are acrocentric and the sexual counterparts submetacentric, was determined for *Canis familiaris* (CHIARELLI 1966; GUSTAVSSON 1964; HARE et al. 1965). This morphology increases the difficulty of chromosome pairing. Thus, most ideograms were initially deduced from more than one karyotype (SELDEN et al., 1975).

With the development of chromosome banding techniques, it was possible to arrive at a more precise description of the canine karyotype. The first description of the G-banding pattern was published by SELDEN et al. (1975). This report established an idiogram with 331 bands per haploid set. A year later, MANOLACHE et al. (1976) presented a representation scheme with 230 G-bands and a different chromosome arrangement. In addition, the authors described the patterns of the Q and C bands of the canine karyotype.

The next G-banding ideogram for canine chromosomes was proposed by Fujinaga et al. (1989). In this ideogram, the arrangement of the chromosomes followed the results presented by Manolache et al. (1976) and the number of G-bands per haploid was 228. This work included the N-band pattern of the canine karyotype. Another ideogram was produced by Stone et al. (1991), using the cell synchronization technique. In the schematic representation they obtained 327 bands, and the arrangement of the chromosomes followed Selden et al. (1975). Graphodatsky et al. (1995) published a canine ideogram with 460 bands and its characteristics were described. The alignment of the chromosomes did not follow any of the previously described works.

Furthermore, the ideograms for the R band have been repeatedly described, each with different chromosome arrangements and band numbers (HOWARD- PEEBLES and PRYOR 1980; POULSEN et al., 1990).

During the *11th European Colloqium on Cytogenetics of Domestic Animals* in 1994, the Canine Chromosome Standardization Committee was established, made up of 6 members from different laboratories. The first step taken by this group was to clarify the question of how many pairs of chromosomes could be unequivocally recognized using G-banding techniques on metaphase chromosomes (REIMANN et al., 1999).

These efforts resulted in the standardization of the G bands of the first 21 autosomes and the sex chromosomes (SWITONSKI et al. 1996). Based on the resolution of bands by this technique, it was noted that the unequivocal identification of the remaining 17 autosomal pairs would require the integration of molecular cytogenetic methods. It was only in 1996 that specific stains for canine chromosomes were developed by Langford et al., 1996.

Thus, the nomenclature used in canine cytogenetic studies follows standards defined by the Committee for the Standardization of the Dog Karyotype (SELDEN et al., 1975; MAYR et al., 1983). However, the nomenclature for describing chromosomal alterations in canine karyotypes has not yet been established. In these cases, it is recommended to use the ISCN (1995) as a reference (REIMANN et al., 1999).

Cytogenetic studies involving conventional staining in mammary carcinoma cells have revealed cases of chromosomal hypodiploidy, hyperdiploidy and tetraploidy (OWEN et al., 1977, 1981; ELSE et al., 1982; NORVALL et al. 1984; WOLFE et al., 1986).

In 1990, a study carried out by Mayr and colleagues cytogenetically characterized a canine hemangiopericytoma using the karyotyping technique (HSU and MOORHEAD, 1956, modified), in which metaphase cells were analyzed. All the tumor cells showed a reduced number of chromosomes (2n=74), and structural alterations such as deletions and centric fusions involving chromosomes 5/6, 5/14, 7/15 and 9/17, as well as a deletion in chromosome 1, were observed.

Cytogenetic studies in two cases of canine osteosarcoma revealed numerical and structural alterations such as several centric fusions identified on chromosomes 1/3, 1/30 and 2/19 in one of the cases and 3/19, 12/18, 13/17 in the other animal (MAYR and ESCHBORN, 1991).

A cytogenetic evaluation of mammary tumors in two domestic dogs (MAYR et al., 1991) found that one of the animals had a normal karyotype in 20% of the metaphases analyzed. In the rest of the cells, a chromosomal marker could be detected, which is the result of a terminal deletion affecting the distal half of the long arm of the X chromosome. The other animal analyzed had trisomy 1 in 20% of the metaphase cells and monosomy X in addition to trisomy 1 in 30% of the metaphase cells.

In 1992, Mayr and Kramberger-Kaplan cytogenetically analyzed canine mammary sarcomas; in one of the cases they found that 60% of the metaphases observed showed a normal karyotype, and in the rest, structural and numerical alterations.

In 1994, Mayr analyzed four solid tumors in dogs and found structural and numerical changes in the karyotype composition of these animals. Metacentric chromosomes, centric fusions and translocations were found.

Chromosomal alterations are associated with factors that control DNA replication and repair, leading to a predisposition to cancer. In many cases, loss of function of tumor suppressor genes (such as p53) and gain of function of oncogenes seem to contribute to the development of tumors (KRAEGEL, 1995, MAYR, 1999).

Through cytogenetic analysis of two canine osteosarcomas, it was possible to detect numerical and structural alterations in all the metaphases observed. The methodology followed protocols for the culture of tumor cells from primary explants and the analysis of metaphases stained by Giemsa and G-banding (WANG AND FEDOROFF, 1972). In one of the cancers, 25 cells were analyzed and all the cells had hyperdiploid chromosome numbers of around 85 to 99. Among the cells investigated, there were 1/3, 1/30, 2/9 centric fusions and structurally altered chromosome 5 (der 5). In the other chamber, 12 cells were analyzed, among which there was a chromosome count of 65 to 67. All the cells showed centric fusions 3/19, 3/?, 12/18 and 13/17 (MAYR et al., 1991).

STONE et al. (1991) cytogenetically evaluated 4 canine mastocytomas. The 4 tumors showed cells with hyperdiploidy and hypodiploidy, an increase in the number of metacentric chromosomes, changes in configuration, and cells showing loss of an X chromosome. One tumor sample showed 15% of the cells evaluated with a chromosome number of 93.

The cytogenetic analysis of canine malignant melanoma showed that half of the metaphases analyzed had normal canine chromosomes (2n=78), with no alterations detected. However, the other metaphases observed had a count of 79 to 81 chromosomes. In these hyperdiploid cells, centric fusions were observed, partially identified as 6/?, 10/? and 11/? (MAYR et al., 1992).

Another study cytogenetically analyzed mammary sarcomas in three dogs, which showed hyperdiploidy, centric fusions and various structural aberrations.

In one of the dogs, 60% of the metaphases analyzed showed normal canine karyotype (2n=78), with no alterations detected. There was hyperdiploidy in 40% of the cells, with chromosome counts of 92 to 98. In the second chamber, none of the 15 metaphases analyzed had a normal canine karyotype. All showed 2n=76 and aberrations such as a double-armed chromosome 6 (6/?), a structurally altered single-armed chromosome 6 (der 6).

In cancer 3, all 30 metaphases analyzed showed numerical alterations, 70% of which were hypodiploid, with chromosome counts ranging from 60 to 66. The identification of centric fusion bands was incomplete due to the very small size of the participating chromosome arms. A total of 30% of the metaphases were described as hyperdiploid, with a chromosome count of 115 to 128. In these hyperdiploid cells, between 25 and 32 double-armed autosomes were detected (MAYR et al., 1992).

In order to improve the prediction of the progression of this neoplasm, there is a constant search for prognostic indicators, and cytogenetics can be considered a tool for diagnosing and predicting the progression of this disease.

4 OBJECTIVES

With the particular aim of contributing to science and the elucidation of new veterinary diagnostic and therapeutic techniques, this work had technical objectives, which are described below.

4.1 GENERAL OBJECTIVES

The main objective of this study is to compare the canine karyotype observed in peripheral lymphocytes with the karyotype obtained from mastocytoma tumor mass cells, in order to observe possible alterations in the chromosomes of Boxer breed dogs with this neoplasm.

4.2 SPECIFIC OBJECTIVES

- To standardize karyotyping techniques for cells from canine mastocytomas.
- To comparatively evaluate the karyotype of peripheral lymphocytes and tumor cells in Boxer dogs.
- Seeking to detect possible numerical and/or structural chromosomal alterations in the karyotypes of tumor tissues.
- Use C and G banding techniques to observe possible chromosomal alterations.
- To identify the Nucleolus Organizing Regions in canine chromosomes using the silver nitrate banding technique.
- Evaluate, if possible, the incidence of chromosomal aberrations and associate them with the degree of malignancy of the tumor.

5 MATERIAL AND METHODS

The procedures used to carry out this work are described in stages below.

5.1 EXPERIMENTAL MODEL

A retrospective study was carried out to establish the prevalence of mastocytomas in relation to dog breeds. The pidemiological survey was carried out at the Animal Pathology Department of the FMVZ - USP Veterinary Hospital, and covered the periods between July 1999 and July 2003 and from August 2004 to April 2008, among all the cases of canine neoplasms treated in the department. Among the cases of mastocytomas, the animals' medical records were consulted. All histopathological examinations were based on the Patnaik classification (1984).

The epidemiological survey followed up all the animals that had material collected.

For this experiment, around 30 dogs of various breeds with mastocytoma were used, with no previous age established and no previous treatment, during the period of standardization of the techniques, until the results of the culture showed sufficient cell numbers to enable the subsequent cytogenetic protocol. As a selection criterion, none of the animals had received previous chemotherapy and/or radiotherapy.

Among the Boxer animals that had their materials collected (blood and tumor tissue), 3 showed good results in the culture protocol and could be analyzed cytogenetically:

Experiment identification number: **38**

Date of collection: **17/05/2007**

Registration: **151626**

Breed: **Boxer**

Name: **Chico**

Sex: **Male**

Birth: **26/10/2000**

Diagnosis: Grade **II** mastocytoma

Experiment identification number: **42**

Date of collection: **12/07/2007**

Registration: **181648**

Breed: **Boxer**

Name: **Naja**

Sex: **Female**

Born: **02/12/1997**

Diagnosis: Grade **III** mastocytoma

Experiment identification number: **26**

Date of collection: **12/09/2007**

Registration: **171469**

Breed: **Boxer**

Name: **Mell**

Sex: **Female**

Birth: **26/06/2000**

Diagnosis: Grade **II** masocytoma

5.2 CLASSIFICATION OF TUMORS

Fine-needle aspiration cytology was carried out on the animals to establish the initial diagnosis of mastocytoma and, after surgical excision, the tumors were sent to the Animal Pathology Service at the Veterinary Hospital (FMVZ-USP) and classified according to their histological grade and staging, based on the Patnaik classification (1984).

5.3 BLOOD COLLECTION

The animals' blood was collected using a sterile 10 mL syringe with a 23-gauge needle and 700 µl of sodium heparin used as an anticoagulant. The blood collected in the syringe was transferred to a sterile *falcon* tube and centrifuged at 1800 rpm for 10 minutes at 37°C for sedimentation of the material and formation of the *buffy coat* with separation of the liquid and cellular parts of the blood. 500 µl of the leukocyte ring was used for the cell culture.

5.4 LEUKOCYTE CELL CULTURE

The cell culture was implanted from 0.5 mL of the leukocyte ring and placed in complete culture medium (RPMI 1640, 0.5% L-glutamine, 0.5% penicillin-streptomycin antibiotic, 20% fetal bovine serum) with 200 µl of phytohemagglutinin-P for 72 hours in an oven at 37°C, 5.0% CO_2. The entire procedure for preparing the leukocyte culture was carried out under completely aseptic conditions in a Trox® laminar flow.

5.4.1 METAPHYSING CHROMOSOMES (HSU and MOORHEAD, 1956, modified)

After 72 hours of incubation, the colchicinization step was carried out. To do this, 100 µl of colchicine [16 µgλmL] was added to each 5 mL of final culture volume, and incubated in an oven at 37°C for 50 minutes.

After colchicinization, the hypotonization step was carried out, where the contents of the culture bottle were transferred to a *falcon* tube and centrifuged at 1800 rpm for 10 minutes at 37°C. The supernatant was carefully discarded using a Pasteur pipette and the *pellet* was resuspended by adjusting the volume of the flask to 7 mL with hypotonic solution (0.075M KCl), carefully added using a Pasteur pipette. The tube was incubated in a water bath at 37°C for 20 minutes. After this period, the tube was removed from the water bath, placed at room temperature, 6 drops of cold fixative (75% methanol and 25% acetic acid) were added and the tube was homogenized and left for a further 10 minutes.

After the hypotonization period, the material was centrifuged at 1800 rpm for 10 minutes to fix it. The supernatant was discarded and the volume of the flask was adjusted to 7 mL with ice-cold fixative. The contents were homogenized and centrifuged again at 1800 rpm for 10 minutes. The supernatant was discarded using a Pasteur pipette and the *pellet* was resuspended with ice-cold fixative, adjusting the volume to 7 mL. This procedure was carried out two more times to remove any remaining cells from the preparation.

5.4.2 PREPARATION OF SLIDES FOR STAINING

After all the stages of processing the material had been carried out, resulting in a *pellet* for cytogenetic preparation, *the* dilution of the material was adjusted with the fixative so that the material was not saturated. The slide used was cleaned with 70% alcohol and a soft diaper. The clean slide was placed on a support in the water bath at 60°C to warm it up. The cytogenetic preparation material was then dripped onto the slide using a Pasteur pipette, at an average height of 30 cm. After the material had dried (24 hours at room temperature), the slides were ready to be submitted to the staining techniques.

5.5 PRIMARY CULTURE OF MASTOCYTOMAS (DEVINNEY and GOLD, 1990)

We used fragments of tumors that had been diagnosed with mastocytoma by the Animal Pathology Service of the Pathology Department (FMVZ-USP) through aspiration cytology from dogs. These materials were obtained from the Surgery Department of the Veterinary Hospital (FMVZ-USP). The fragments were kept in RPMI culture medium at room temperature for a maximum of 1 hour before the cultures were prepared. In the laboratory, under aseptic conditions, the fragments of canine mastocytomas were mechanically broken into pieces of approximately 1mm and incubated in a solution of collagenase type IA from

Clostridium histolyticum (100U/ml) in RPMI and antibiotics for about 2 hours, at 37°C and in a shaking water bath.

The resulting cell suspension was filtered through gauze and then through a 70 µm filter. The entire contents were centrifuged at 1200 rpm for 10 minutes and the supernatant was discarded. The sedimented cells were resuspended in 10 ml of RPMI supplemented with 10% Fetal Bovine Serum, antibiotics and placed in culture in an incubator at 37°C, 5.0% CO_2. The culture bottle was observed daily until the cells showed a high rate of division, which occurred after 96 hours of incubation (Figure 18).

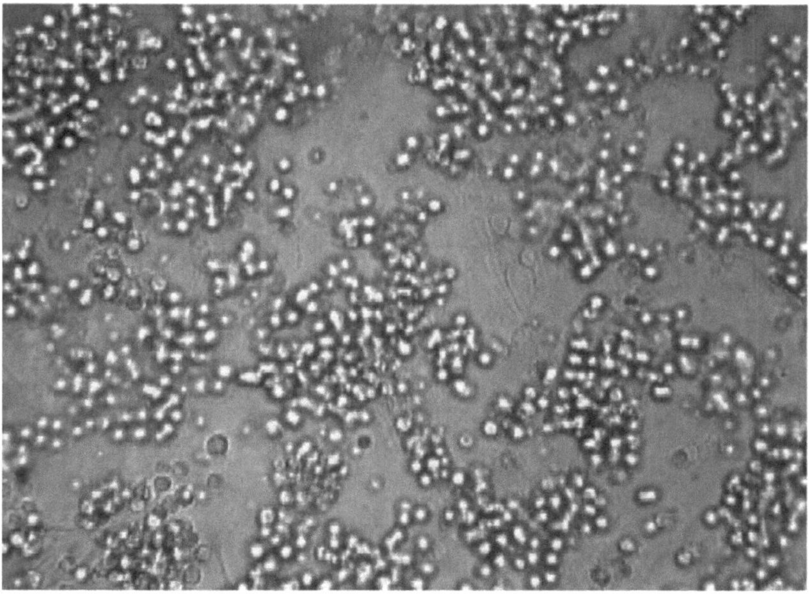

Figure 18 - Mastocytoma tumor tissue in culture. Presence of mast cells and fibroblasts. Viewed under a Motic inverted microscope (200x).

5.5.1 PROCESSING FOR CHROMOSOME ACQUISITION IN METAPHASE (HSU and MOORHEAD, 1956, modified)

After the cell culture incubation period, the colchicinization step was carried out. To do this, 100 µl of colchicine [16 µg/mL] was added to each 5 mL of final culture volume, and incubated in an oven at 37°c for 50 minutes.

After the colchicinization stage, the cultured cells were trypsinized. At this stage, the suspension from the culture bottle was removed and stored in a *falcon* tube. 3 mL of Hank's solution (without calcium or magnesium) was added to the bottle and incubated in the oven for 20 minutes. The Hank's solution was then removed from the culture bottle and stored together with the culture medium removed in the previous step. A quantity of 2 mL of trypsin EDTA was added to the culture bottle, which was shaken slightly so that the adhered cells were detached from the wall of the culture bottle. This procedure lasts no longer than 1 minute. The trypsin solution and the detached cells were added to the previously removed contents and everything was centrifuged at 1800 rpm for 10 minutes. The supernatant was discarded using a Pasteur pipette.

After trypsinization, the hypotonization step was carried out. The entire contents of the tube were transferred to a *falcon* tube and centrifuged at 1800 rpm for 10 minutes at 37°C. The supernatant was carefully discarded using a Pasteur pipette and the *pellet* was resuspended by adjusting the volume of the flask to 7 mL with hypotonic solution (0.075M KCl), carefully added using a Pasteur pipette. The contents of the tube were incubated in a water bath at 37°C for 20 minutes. After this period, the tube was removed from the water bath and placed at room temperature. 6 drops of ice-cold fixative (75% methanol and 25% acetic acid) were added to the contents and homogenized for a further 10 minutes.

After the hypotonization period, the material was centrifuged at 1800 rpm for 10 minutes to fix it. The supernatant was discarded and the volume of the flask was adjusted to 7 mL with ice-cold fixative. The contents were homogenized and centrifuged again at 1800 rpm for 10 minutes. The supernatant was discarded using a Pasteur pipette and the *pellet* was resuspended by adjusting the volume to 7 mL with ice-cold fixative, added carefully. This procedure was carried out two more times to remove any remaining cells from the preparation.

5.5.2 PREPARATION OF SLIDES FOR STAINING

After obtaining a cytogenetic preparation *pellet,* the dilution of the material was adjusted with the fixative so that the material was not saturated. The slide used was cleaned with 70% alcohol and a soft diaper. The cleaned slide was placed on a support in the water bath at 60°C to warm it up. The cytogenetic preparation material was then dripped onto the slide using a Pasteur pipette, at an average height of 30 cm. After the material had dried (24 hours at room temperature), the slides were ready to be submitted to the staining techniques.

5.6 COLORING TECHNIQUES

5.6.1 CONVENTIONAL COLORING

The slides were hydrolyzed in a 1N HCl solution, bathed for 7 minutes at 60°C and washed with distilled water. After drying for a few minutes, the slides were stained with Giemsa (1:30) in phosphate buffer for 7 minutes and washed with distilled water. After drying, the slides were submitted for analysis.

5.6.2 BAND C (CBG), according to Sumner (1972), with modifications

The slides were hydrolyzed in 0.2N HCl for 30 minutes at room temperature and washed with distilled water. After drying for a few minutes, the slides were dipped in 5% barium hydroxide octahydrate at 60°C for approximately 10 seconds. Immediately afterwards, the slides were rinsed with distilled water and quickly dipped in 1N HCl at 60°C. After this procedure, they were washed in distilled water and dried. The slides were then incubated in 2xSSC solution (pH 7.0) at 60°C for 15 minutes, washed in distilled water and dried. The slides were stained with Giemsa (1:20) in phosphate buffer for 30 minutes and washed with distilled water. After drying, the slides were submitted for analysis.

5.6.3 G-BAND (GTG), according to Seabright (1971), with modifications

The slides were hydrolyzed for 15 minutes in 2xSSC at 60°C and washed with distilled water. After a few minutes of drying, the slides were immersed in a trypsin solution (0.02g in 75 ml phosphate buffer pH 6.8) at 37°C, moving back and forth for approximately 8 seconds. After this procedure, they were washed in distilled water, dried and stained with Giemsa (1:30) in phosphate buffer for 7 minutes and washed with distilled water. After drying, the slides were submitted for analysis.

5.6.4 RON BAND (NUCLELEOL ORGANIZING REGIONS), according to Howell and Black (1980), with modifications

The slides were hydrolyzed in 1N HCl at 60°C for 7 minutes at room temperature and washed with distilled water. After a few minutes of drying, 1 drop of colloidal developer solution (1g of Merck® elatin for microbiology in 50 ml of distilled water with 0.5 ml of formic acid) and two drops of 50% silver nitrate solution ($AgNO_3$) were dripped onto the slide material. The coverslips were covered and incubated in a humid chamber at 60°C for about 3 to 4 minutes. After this period, the coverslips were removed, the slides were washed with distilled water and stained with Giemsa solution (Merck®) 1:50 for 30 seconds.

5.7 METAPHASE ANALYSIS

After the slides had dried completely, the material was analyzed and photographed using a Zeiss® immersion microscope with a 100x objective and a green filter, equipped with Ikaros® cytogenetic analysis software. Numerical and structural changes were considered.

Complete metaphases for analysis were those with material well spread out on the slide and with good visualization, which implies the condensation of the genetic material and can be observed later, in the chapter of results.

All the metaphases photographed were identified according to their location on the slide (microscope coordinate), slide number and metaphase number, for control, traceability and data reliability.

The chromosomes were computed with the help of a specific tool in the cytogenetic analysis system to investigate numerical changes, and were also observed to look for structural changes.

The metaphases of peripheral lymphocytes and tumor tissue cells were analyzed using conventional staining (Giemsa), G bands, C bands and Rons bands.

6 RESULTS

This stage describes and justifies the results of all the procedures carried out in this work.

6.1 RETROSPECTIVE STUDIES

The retrospective study carried out on the incidence of mastocytomas in dogs between July 1999 and July 2003 analyzed the histopathological reports obtained at the Pathological Anatomy Department, FMVZ-USP, which were evaluated. Of the total of 4090 histopathological examinations carried out in the period, 2.3% of the reports (94) were related to cases of mastocytomas in dogs, of which 53 animals (56.4%) were male. Ages ranged from one to 15 years, with an average of 8.13 ± 3.02; 58.6% of the dogs were between seven and ten years old.

There was a higher prevalence of mastocytomas in animals without a defined breed (37.2%), Boxers (27.7%) and Poodles (7.4%). In terms of tumor distribution, the most affected skin regions were the scrotum (25.5%) and breast (7.4%). Of the tumors analyzed, 14% had compromised surgical margins, 6.4% occurred together with seminoma, 5.3% with leydigoma and 3.2% with hemangioma.

According to Patnaik's classification (1984), 16% of the tumors had grade I, 26.6% grade II, 16% grade III and 4.3% had grades II and III in the same animal. A total of 37.2% of mastocytomas were not classified according to the grade described in the medical records.

A survey of data from histopathological examinations carried out by the Animal Pathology Service at HOVET - FMVZ - USP, between August 2004 and April 2008, revealed 206 cases of mastocytomas in dogs.

With regard to breed, 31% of the cases occurred in Boxer breed animals, followed by 24% of animals with no defined breed, 8% of Labradors and the remainder (37%) corresponding to the total of other breeds, which does not correspond to a pure breed.

The age of the dogs ranged from three to 15 years, with a prevalence of 42% of animals between seven and ten years old. Female animals accounted for 53% of the cases. According to the histopathological grading based on Patnaik (1984), the majority of cases, represented by a total of 56%, were diagnosed with grade II mastocytoma, followed by 23% of animals with grade I mastocytoma and 12% diagnosed with grade III mastocytoma. There was a small percentage of animals affected by more than one type of histopathological grade of mastocytoma.

6.2 STANDARDIZATION OF TECHNIQUES

Materials from dogs of different breeds were included in the experiment until the techniques were standardized. Once the culture techniques and cytogenetic protocol were standardized, only material from Boxer breed dogs was collected.

Tests were carried out to standardize the cell culture technique for use in cytogenetic protocols. Three time periods (48, 72 and 96 hours) were evaluated for incubation in an incubator at 37°C, 5.0% CO_2. The period that showed the best results in cell division was 72 hours for peripheral lymphocyte culture and 96 hours for primary tumor tissue culture.

Once the incubation period had been standardized, steps in the protocol for karyotyping the cells resulting from the culture were analyzed. To establish the KCl concentration for cell lysis, methods from other published works were compared. Two concentrations of the reagent were tested: 0.075M and 0.04M, and the first provided permeability of the cells without them losing their chromosome content (Figure 19).

The incubation period of the culture material with colchicine was also well defined, as the chromosomes would appear with separated chromatids if the time was extended (Figure 20), even by a few minutes. This time was then standardized at 50 minutes for tumor cell cultures and 60 minutes for peripheral lymphocyte cultures.

It was necessary to set a height to drip the material onto the slide so that it could spread over the surface. For this procedure, the slides were placed on a rack inside the water bath and the material was dripped using a Pasteur pipette. In the first experiments, the material was dripped at an average height of 20 cm, which meant that the metaphases were observed very close together, not allowing for a reliable analysis. After a few tests, it was observed that the best height for depositing the material on the slide is between 35 and 40 cm.

The staining tests were established in relation to the reaction times with the solutions. For example, the standardized period for the trypsin bath, for G banding, was nine seconds in motion, at 60°C. For the 5% barium bath for C banding, a period of 25 seconds in motion at 60°C was established. The period of contact between the slides and the silver nitrate solution was standardized at three minutes and 15 seconds.

While the reaction times with the solutions were standardized, different results were observed, such as excessive corrosion of the chromosomal material when the time in contact with the substance was exceeded.

6.3 THE ANIMALS

Among the previously selected Boxer breed animals, we obtained good quality tumor material and cell division from three animals, which were submitted to the cell culture and cytogenetics protocol.

After the surgical procedure, the animals analyzed were followed up for treatment and survival time.

Chico had a mastocytoma in the hind limb, with tumor formation in the subcutaneous tissue, a 30-month evolution and a 12-month interval for recurrence. The animal did not have compromised surgical margins or metastasis, and was grade II according to the Patnaik classification (1984). The animal is still alive.

The dog Naja had a mastocytoma on her hind limb and head, with nine tumor formations, a 24-month evolution and a dermal spread. The animal did not have compromised surgical margins or metastasis, and had tumor grade III according to the Patnaik classification (1984). The dog survived 12 months after surgery and died.

The dog Mell had a mastocytoma on her hind limb and back, with two tumor formations, a two-month evolution and a dermal spread. The animal did not have compromised surgical margins or metastasis; it was grade II according to Patnaik's classification (1984). There was a five-month period of tumor recurrence, which was being controlled until the animal contracted leptospirosis, which worsened its clinical condition due to weakness, and it died.

6.4 GENETIC ALTERATIONS

A comparative study was carried out to evaluate the number of chromosomes, analyzing metaphases of peripheral lymphocytes and metaphases of tumor tissue cells in five dogs of the breeds Boxer, Pug, Labrador and an animal without a defined breed. The age of the dogs ranged from seven to 12 years and all the tumors were classified as grade II mastocytoma.

After the cytogenetic preparations, 20 cells from each type of material from each animal were analyzed. In the analysis carried out, the chromosome count in the metaphases of peripheral lymphocytes was unchanged, maintaining the diploid number of 2n=78; however, there was a variation between 74 and 79 chromosomes observed in the material from the animals' tumor cells.

6.5 THE BOXER BREED AND CYTOGENETIC CHANGES

The results of the cytogenetic analysis described here refer to the specific analysis of the material from the three Boxer dogs mentioned above. From all the material collected from the dogs, it was possible to make slides with cytogenetic material from peripheral lymphocytes and tumor cells.

A total of 828 metaphases were analyzed, partially divided for analysis into conventional staining, G bands, C bands and Rons bands (Tables 1, 2, 3 and 4). From peripheral lymphocytes, 476 metaphases were photographed and analyzed. A total of 352 metaphases were analyzed from the animals' tumor material.

Conventional staining analysis						
Animal	Met.	2n=78	2n=77	2n=79	2n=76	2n=75
26S	74	74	0	0	0	0
26T	79	58	14	1	6	2
38S	67	67	0	0	0	0
38T	102	64	22	1	15	0
42S	94	93	1	0	0	0
42T	52	37	9	0	6	0

Table 1 - Met: number of metaphases analyzed; 26S: animal 26, peripheral lymphocytes; 26T: animal 26, tumor cells; 38S: animal 38, peripheral lymphocytes; 38T: animal 38, tumor cells; 42S: animal 42, peripheral lymphocytes; 42T: animal 42, tumor cells

C-banding analysis					
Animal	Met.	2n=78	2n=77	2n=79	2n=76
26S	25	25	0	0	0
26T	10	9	0	0	1
38S	16	16	0	0	0
38T	10	9	1	0	0
42S	22	22	0	0	0
42T	0	0	0	0	0

Table 2 - Met: number of metaphases analyzed; 26S: animal 26, peripheral lymphocytes; 26T: animal 26, tumor cells; 38S: animal 38, peripheral lymphocytes; 38T: animal 38, tumor cells; 42S: animal 42, peripheral lymphocytes; 42T: animal 42, tumor cells

Ron Banding Analysis					
Animal	Met.	2n=78	2n=77	2n=79	2n=76

26S	30	30	0	0	0
26T	30	21	7	1	1
38S	20	20	0	0	0
38T	20	13	5	0	2
42S	20	20	0	0	0
42T	9	6	1	0	2

Table 3 - Met: number of metaphases analyzed; 26S: animal 26, peripheral lymphocytes; 26T: animal 26, tumor cells; 38S: animal 38, peripheral lymphocytes; 38T: animal 38, tumor cells; 42S: animal 42, peripheral lymphocytes; 42T: animal 42, tumor cells

G-banding analysis					
Animal	Met.	2n=78	2n=77	2n=79	2n=76
26S	68	68	0	0	0
26T	20	17	1	1	1
38S	20	20	0	0	0
38T	20	16	2	1	1
42S	20	19	1	0	0
42T	0	0	0	0	0

Table 4 - Met: number of metaphases analyzed; 26S: animal 26, peripheral lymphocytes; 26T: animal 26, tumor cells; 38S: animal 38, peripheral lymphocytes; 38T: animal 38, tumor cells; 42S: animal 42, peripheral lymphocytes; 42T: animal 42, tumor cells

Analysis of metaphases from cell cultures of peripheral lymphocytes at showed no structural alterations. Numerical analysis showed one metaphase with 2n=77, and all the others showed the usual canine diploid number.

In the analysis of tumor cell metaphases, among all the stains, the first alterations were observed. Among the 352 metaphases analyzed, 70.9% showed a normal diploid number, 2n=78. Another 17.4% of the metaphases showed 2n=77; 9.8% had 2n=76. Five metaphases (1.3%) had 2n=79 and two cells had 2n=75. Some analysis results are included in tables 5 and 6.

Color type.	**LP**	**T**	**Total**

Conventional	235	233	468
Band G	108	40	148
Band C	63	20	83
Ron Band	70	59	129
Total	476	352	828

Table 5 - Number of metaphases analyzed by type of staining. LP: metaphases of peripheral lymphocytes; T: metaphases of tumor cells

2n	T	%
78	250	70,9
77	62	17,4
76	35	9,8
79	5	1,3
75	2	0,6
Total	352	100

Table 6 - Incidence of metaphases by diploid number

Structural alterations were also found in the analyses carried out on the metaphase chromosomes of tumor cell cultures from the three animals.

There were metacentric and submetacentric chromosomes, as well as sex chromosomes, and more stained regions in certain parts of the chromosomes.

With regard to the incidence of chromosomal aberrations associated with the degree of malignancy of the tumors, there was no significant difference in the analysis of the metaphases. The three animals showed similar numerical and structural alterations.

Silver nitrate staining marked the end of the chromosomes in the centromeric region, allowing the nucleolus-organizing regions to be visualized.

The following pages contain pictures of some of the changes found.

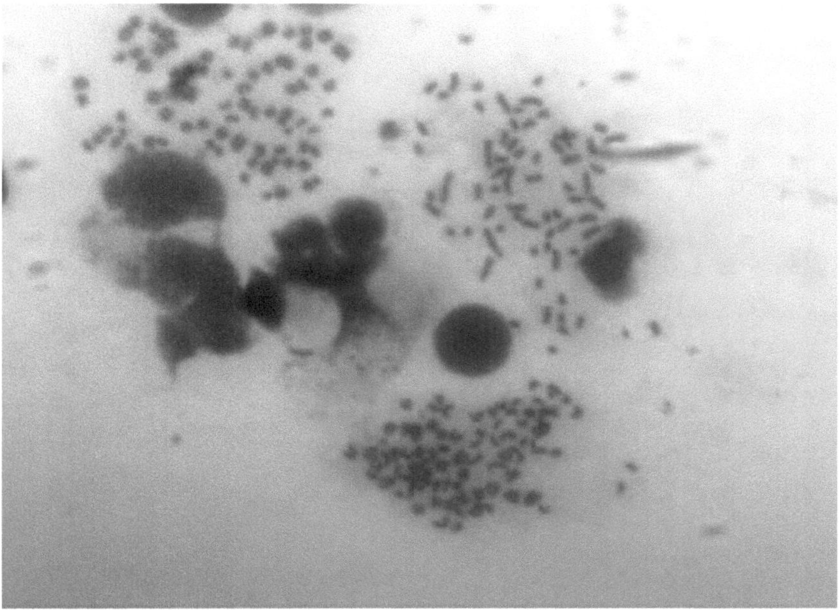

Figura 19 - Metaphase of tumor tissue. Female SRD dog. Observation of 3 cells with chromosomes in very close metaphases, which does not allow analysis due to the possibility of confusion between the chromosomes.

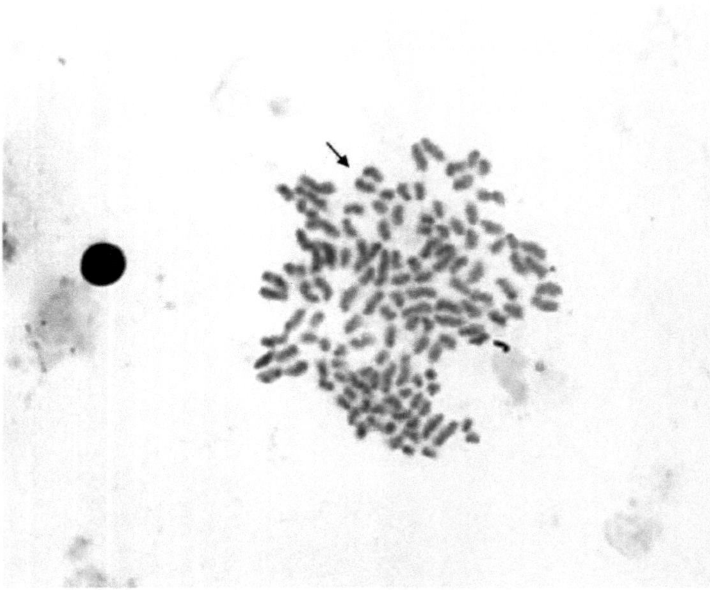

Figura 20 Metaphase of tumor tissue. Male Boxer dog. Separated chromosome arms (arrow), the result of too much colchicine. Viewed under a Zeiss microscope (1000x) and green filter. After a few tests, the colchicine action time was standardized at 50 minutes in a 37°C oven with 5% CO_2, so that the chromosomes remained intact.

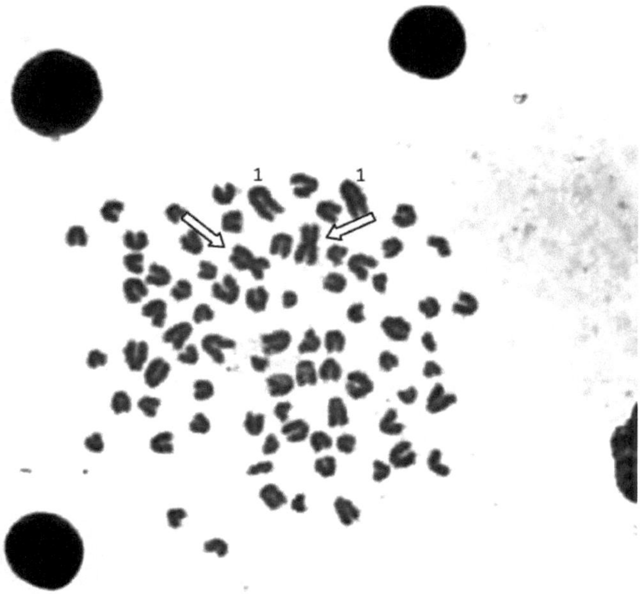

Figura 21 - Canine mitotic metaphase obtained from peripheral lymphocyte culture. Female Boxer dog (2n=78). The wide white arrows indicate the pair of XX sex chromosomes and the first pair is indicated with the number 1. Conventional staining. Viewed under Zeiss microscope (1000x) and green filter

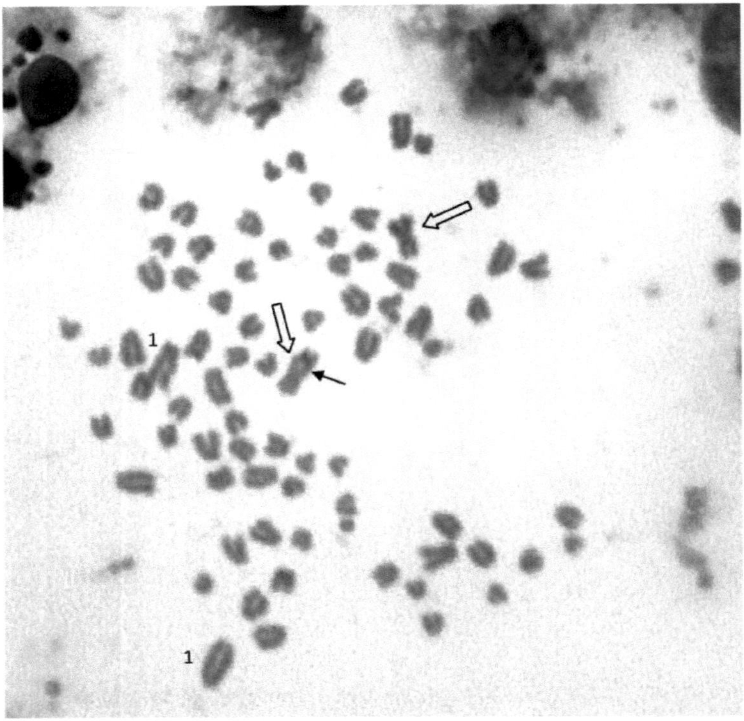

Figure 22 - Canine mitotic metaphase obtained from tumor tissue cell culture. Female Boxer dog (2n=78). The wide white arrows indicate the pair of XX sex chromosomes and the first pair is indicated with the number 1. Observation of heterochromatic marking in the distal region of the long arm of the X chromosome (single arrow). Conventional staining. Visualization under Zeiss microscope (1000x) and green filter

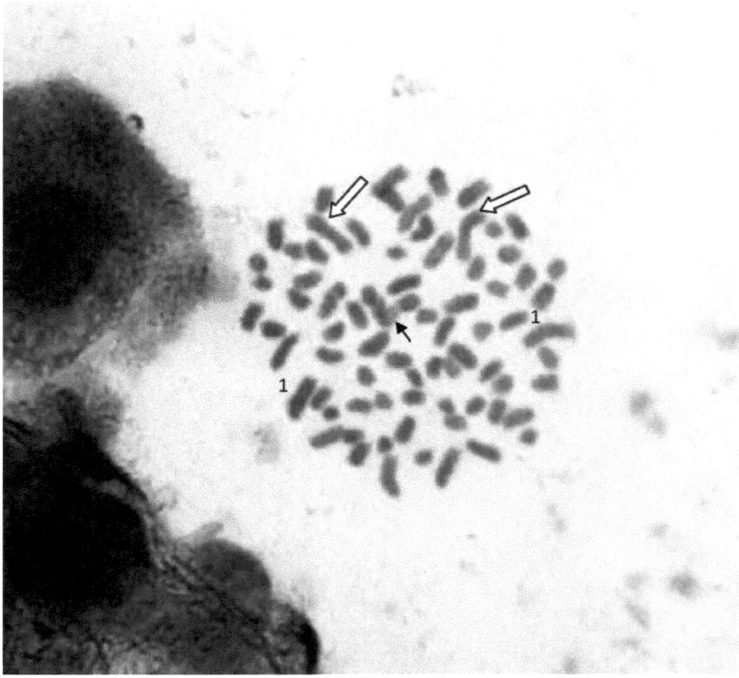

Figure 23 - Canine mitotic metaphase obtained from tumor tissue cell culture. Female Boxer dog (2n=78). The wide white arrows indicate the pair of XX sex chromosomes and the first pair is indicated with the number 1. *HSR* (homogeneously staining region) observation (single arrow). Conventional staining. Visualization under Zeiss microscope (1000x) and green filter

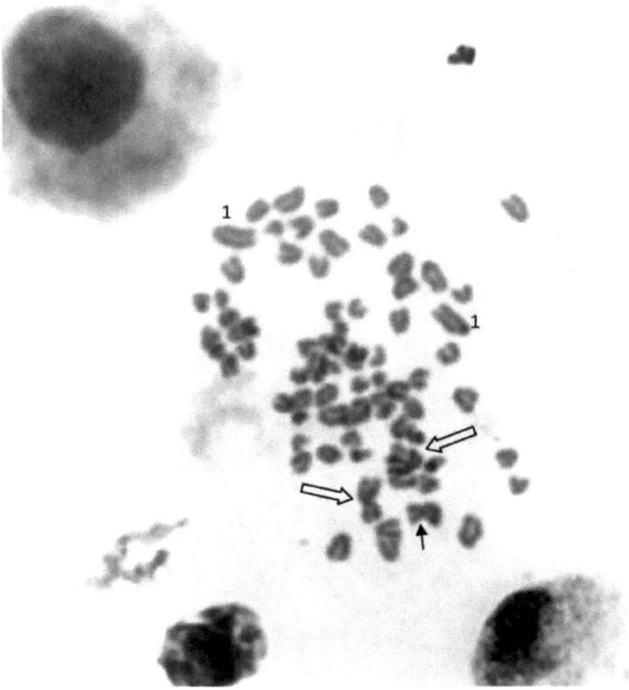

Figure 24 - Canine mitotic metaphase obtained from tumor tissue cell culture. Female Boxer dog (2n=77). Observation of 1 median submetacentric chromosome (single arrow), the result of centric fusion. The wide white arrows indicate the pair of XX sex chromosomes and the first pair is indicated with the number 1. Conventional staining. Viewed under Zeiss microscope (1000x) and green filter

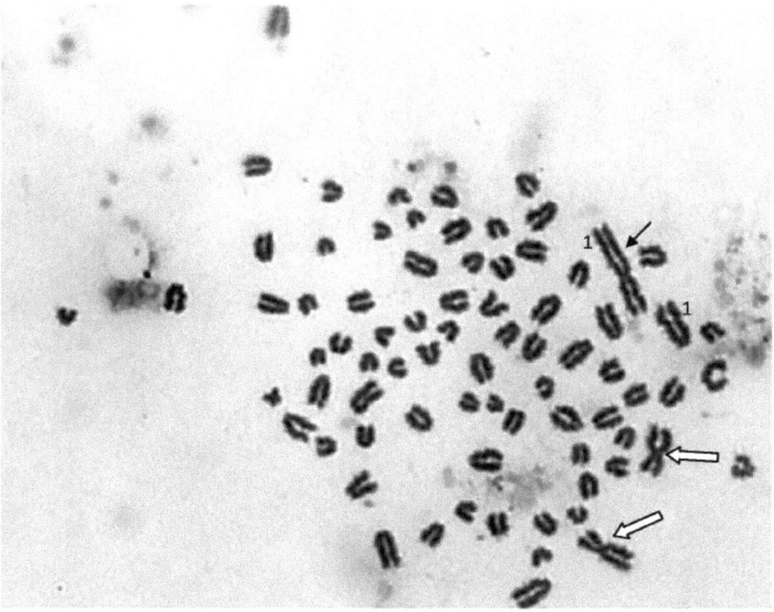

Figure 25 - Canine mitotic metaphase obtained from tumor tissue cell culture. Female Boxer dog (2n=77). Observation of a large submetacentric chromosome (single arrow), the result of centric fusion between a homologue of chromosome number 1 and another acrocentric chromosome. The wide white arrows indicate the pair of XX sex chromosomes and the first pair is indicated with the number 1. Conventional staining. Viewed under Zeiss microscope (1000x) and green filter

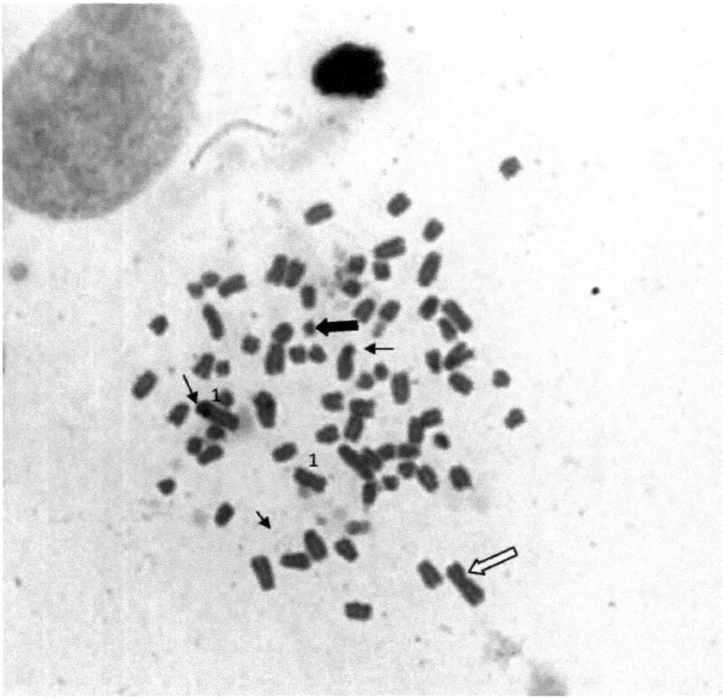

Figure 26 - Canine mitotic metaphase obtained from tumor tissue cell culture. Male Boxer dog (2n=78). Observation of more stained regions in the centromeric region of one of the homologues of pair 1 and in two other chromosomes (single arrows). The wide white arrow indicates the X sex chromosome and the wide black arrow indicates the Y sex chromosome; the first pair is indicated with the number 1. Conventional staining. Viewed under Zeiss microscope (1000x) and green filter

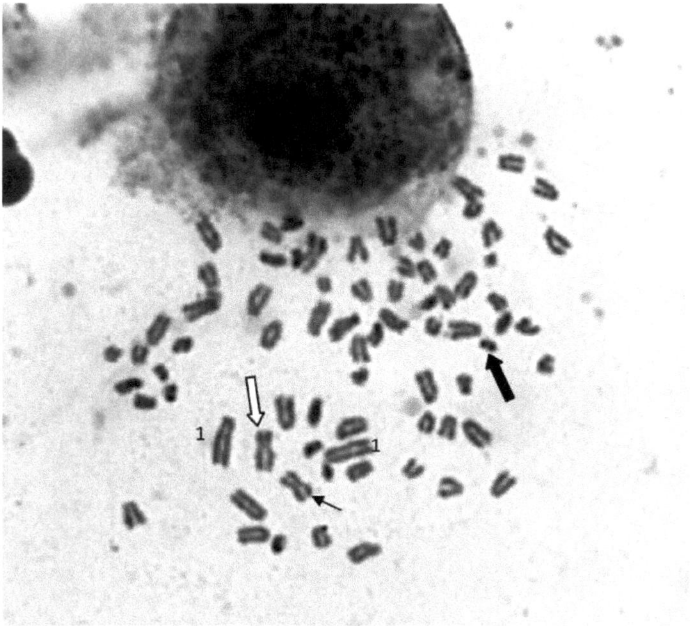

Figure 27 - Canine mitotic metaphase obtained from tumor tissue cell culture. Male Boxer dog (2n=77). Observation of submetacentric chromosome (single arrow) resulting from centric fusion. The wide white arrow indicates the X sex chromosome and the wide black arrow indicates the Y sex chromosome; the first pair is indicated with the number 1. Conventional staining. Viewed under Zeiss microscope (1000x) and green filter

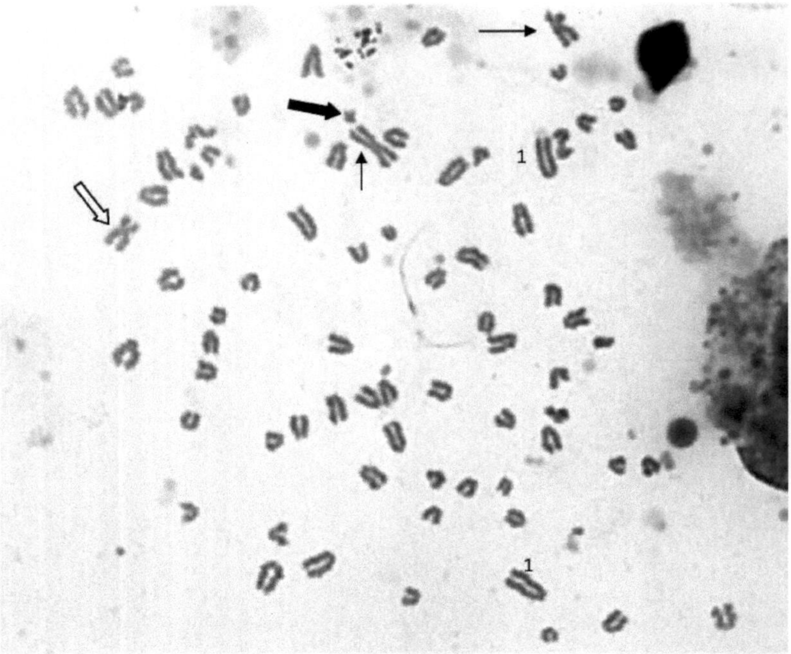

Figure 28 - Canine mitotic metaphase obtained from tumor tissue cell culture. Male Boxer dog (2n=76). Observation of two submetacentric chromosomes (single arrows) resulting from centric fusion. The wide white arrow indicates the X sex chromosome and the wide black arrow indicates the Y sex chromosome; the first pair is indicated with the number 1. Conventional staining. View under Zeiss microscope (1000x) and green filter

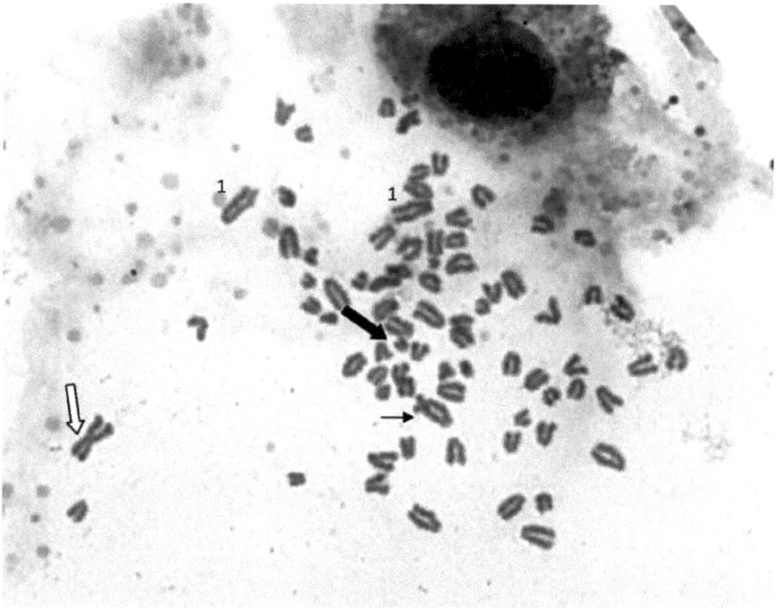

Figure 29 - Canine mitotic metaphase obtained from tumor tissue cell culture. Male Boxer dog (2n=77). Observation of a submetacentric chromosome (single arrow) resulting from centric fusion. The wide white arrow indicates the X sex chromosome and the wide black arrow indicates the Y sex chromosome; the first pair is indicated with the number 1. Conventional staining. View under Zeiss microscope (1000x) and green filter

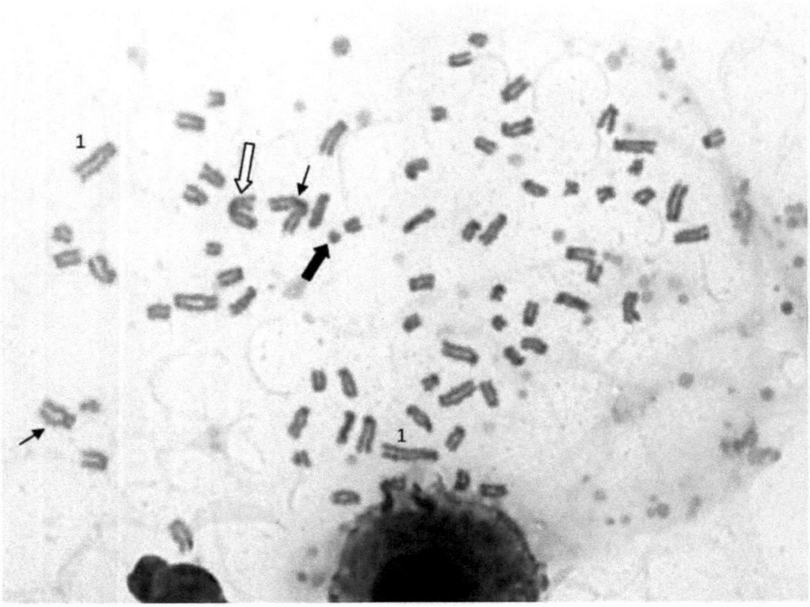

Figure 30 - Canine mitotic metaphase obtained from tumor tissue cell culture. Male Boxer dog (2n=76). Observation of two submetacentric chromosomes (single arrows) resulting from centric fusions. There is also more dye absorption in the centromeres of some chromosomes. The wide white arrow indicates the X sex chromosome and the wide black arrow indicates the Y sex chromosome; the first pair is indicated with the number 1. Conventional staining. View under Zeiss microscope (1000x) and green filter

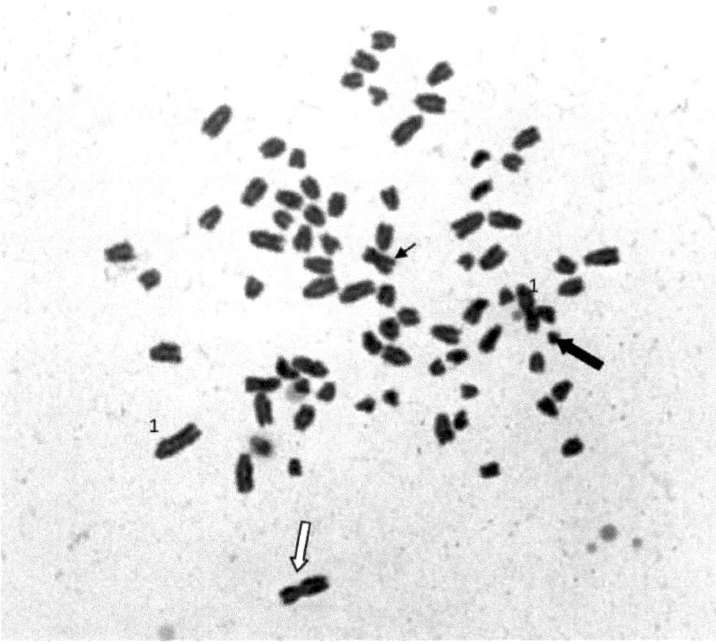

Figure 31 - Canine mitotic metaphase obtained from tumor tissue cell culture. Male Boxer dog (2n=77). Observation of a submetacentric chromosome (single arrow) resulting from centric fusion. The wide white arrow indicates the X sex chromosome and the wide black arrow indicates the Y sex chromosome; the first pair is indicated with the number 1. Conventional staining. View under Zeiss microscope (1000x) and green filter

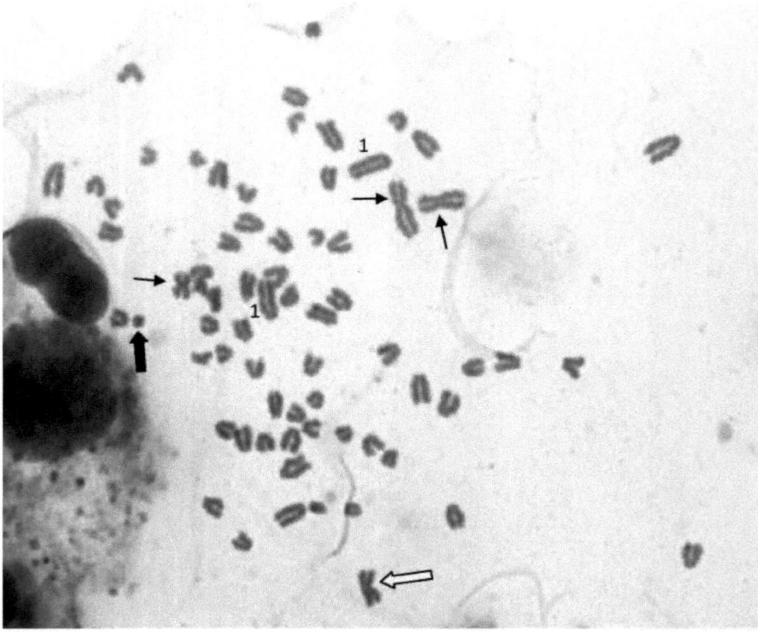

Figure 32 - Canine mitotic metaphase obtained from tumor tissue cell culture. Male Boxer dog (2n=75). Observation of two metacentric chromosomes (one medium and one small) and one large submetacentric chromosome (single arrows) resulting from centric fusions. The wide white arrow indicates the X sex chromosome and the wide black arrow indicates the Y sex chromosome; the first pair is indicated with the number 1. Conventional staining. Viewed under Zeiss microscope (1000x) and green filter

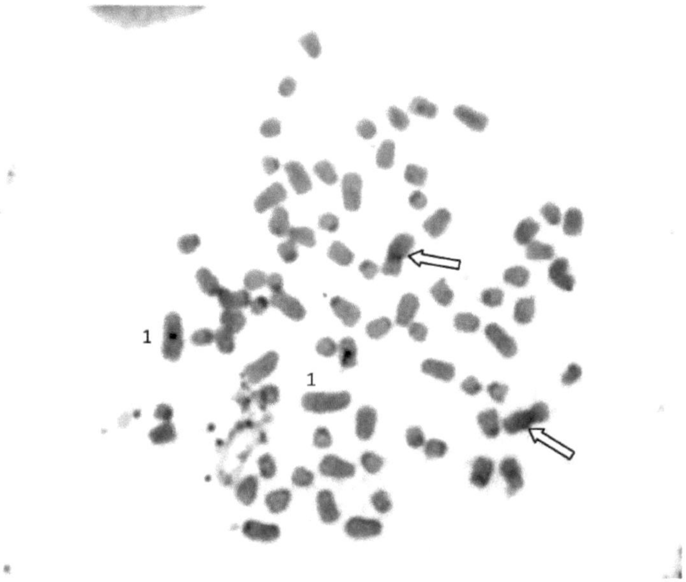

Figure 33 - Canine mitotic metaphase obtained from tumor tissue cell culture. Female Boxer dog (2n=78). Indicated: sexual pair (wide white arrows) and pair number 1. Band C staining. Viewed under Zeiss microscope (1000x) and green filter.

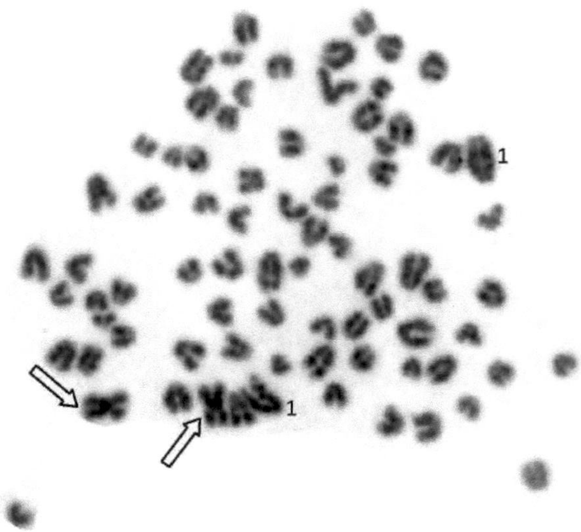

Figure 34 - Metaphase of tumor tissue. Female Boxer dog (2n=78). *HSR* (homogeneously staining region) observation (single arrow). Indicated: sexual pair (wide white arrows) and homologous pair number 1. G-band staining. Viewed under a Zeiss microscope (1000x) and green filter.

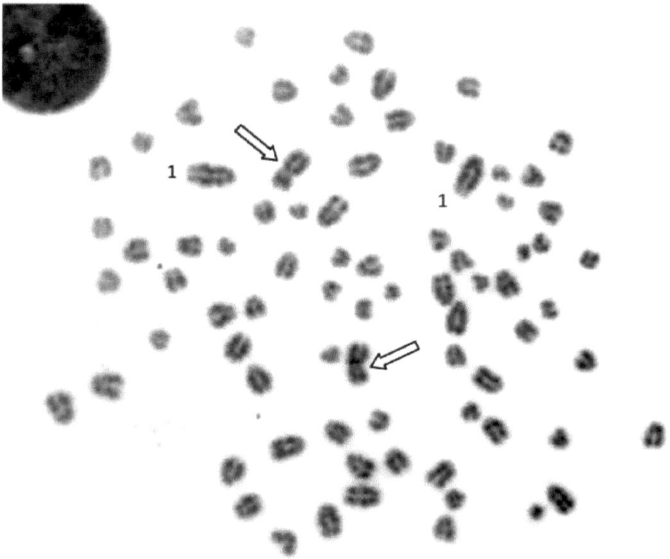

Figure 35 - Canine mitotic metaphase obtained from a culture of peripheral lymphocytes. Female Boxer dog (2n=78). The wide white arrows indicate the pair of XX sex chromosomes and the first pair is identified with the number 1. G-band staining. Viewed under a Zeiss microscope (1000x) and green filter.

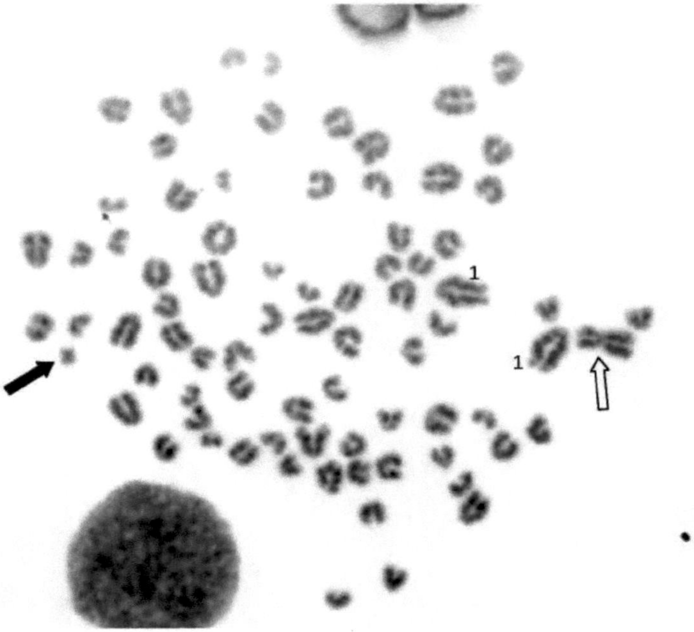

Figure 36 - Canine mitotic metaphase obtained from a culture of peripheral lymphocytes. Male Boxer dog (2n=78). The wide white arrow indicates the X sex chromosome and the wide black arrow indicates the Y sex chromosome; the first pair is indicated with the number 1. Band G staining. Viewed under Zeiss microscope (1000x) and green filter

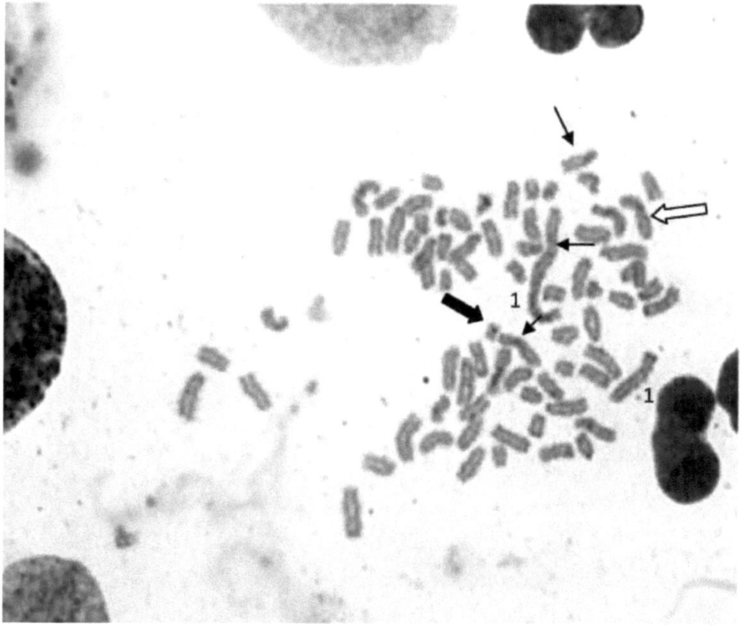

Figure 37 - Metaphase of tumor tissue. Male Boxer dog (2n=75). Observation of three submetacentric chromosomes (single arrows) due to centric fusions. One of the homologues of pair 1 is involved in the rearrangement of the largest submetacentric chromosome found. The wide white arrow indicates the X sex chromosome and the wide black arrow indicates the Y sex chromosome; the first pair is indicated with the number 1. Conventional staining. Viewed under Zeiss microscope (1000x) and green filter

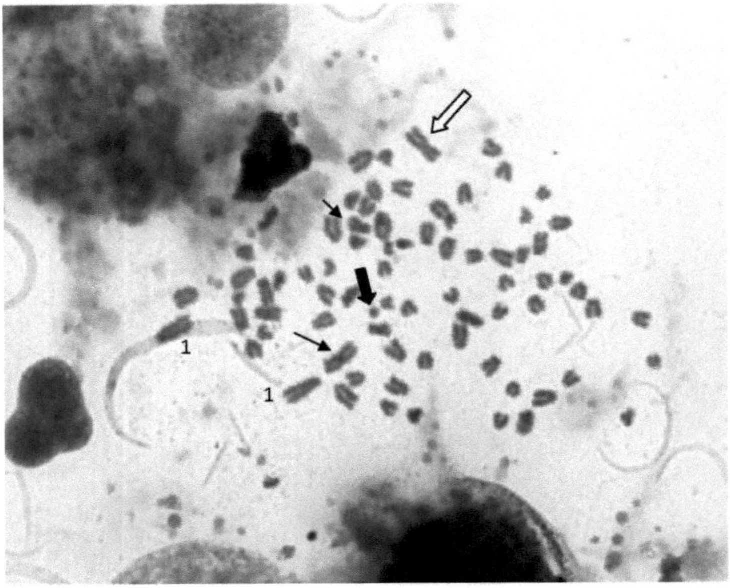

Figure 38 - Metaphase of tumor tissue. Male Boxer dog (2n=76). Observation of a large metacentric chromosome and a small submetacentric chromosome (single arrows) resulting from centric fusions. The wide white arrow indicates the X sex chromosome and the wide black arrow indicates the Y sex chromosome; the first pair is indicated with the number 1. Conventional staining. Viewed under Zeiss microscope (1000x) and green filter

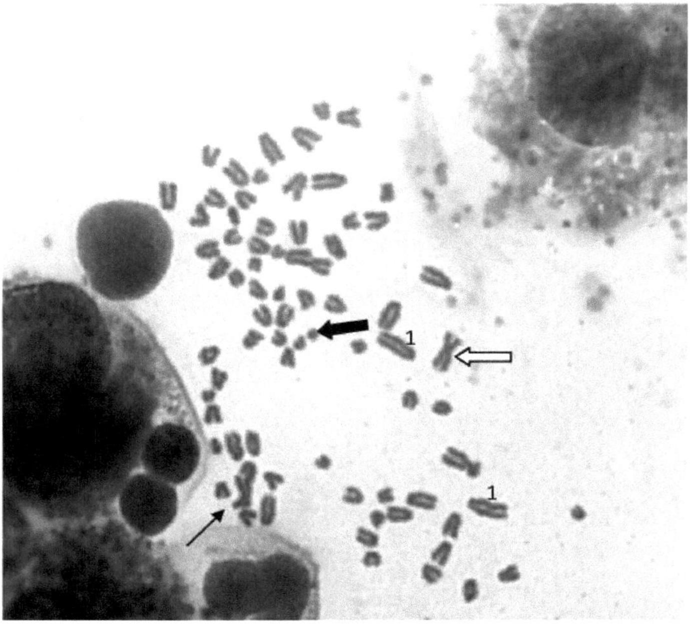

Figure 39 - Metaphase of tumor tissue. Male Boxer dog (2n=77). Observation of a large submetacentric chromosome (single arrows) resulting from centric fusions. The wide white arrow indicates the X sex chromosome and the wide black arrow indicates the Y sex chromosome; the first pair is indicated with the number 1. Conventional staining. Viewed under Zeiss microscope (1000x) and green filter

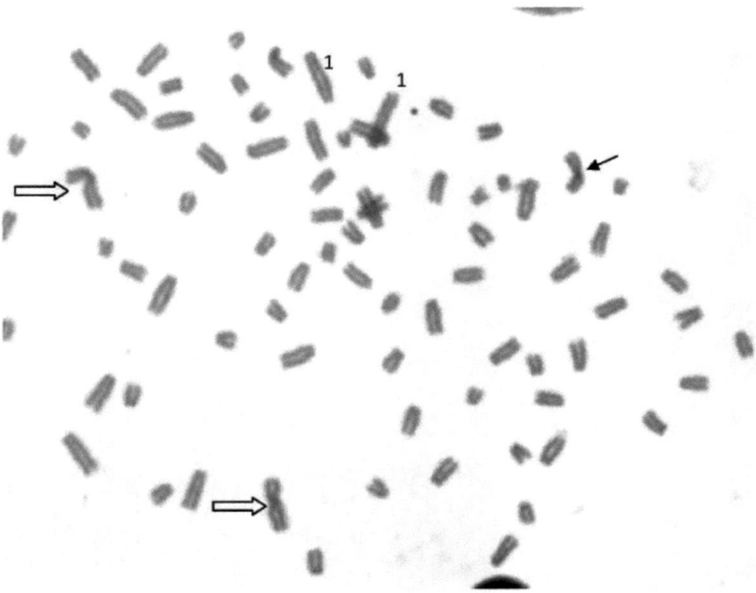

Figure 40 - Metaphase of tumor tissue. Female Boxer dog (2n=77). Observation of 1 submetacentric chromosome (single arrow) al. Indicated: sex pair (wide white arrows) and homologous pair number 1. Conventional staining. Visualization under Zeiss microscope (1000x) and green filter

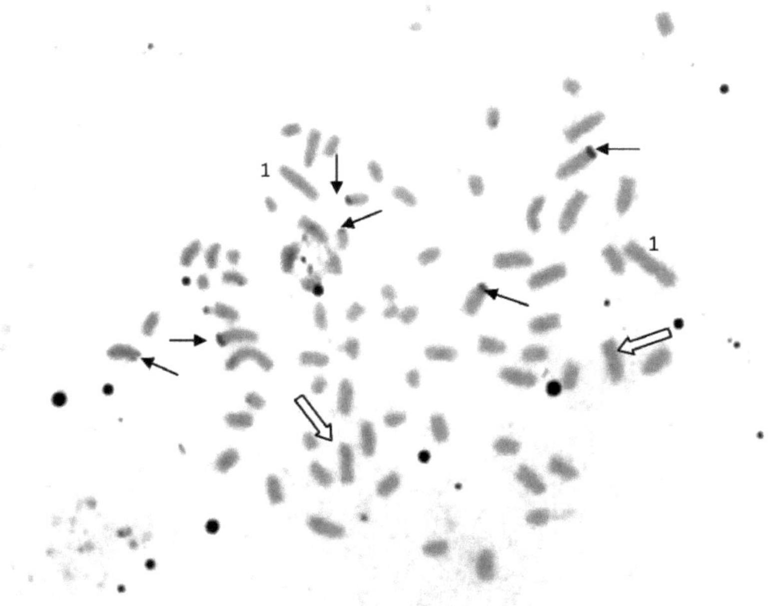

Figure 41 - Metaphase of tumor tissue. Female Boxer dog (2n=78). Observation of silver staining in the telomeric region of six chromosomes (single arrows). Indicated: 1 sexual homologue (wide white arrow) and pair number 1. Silver nitrate staining (Ron Band). Visualization under Zeiss microscope (1000x) and green filter

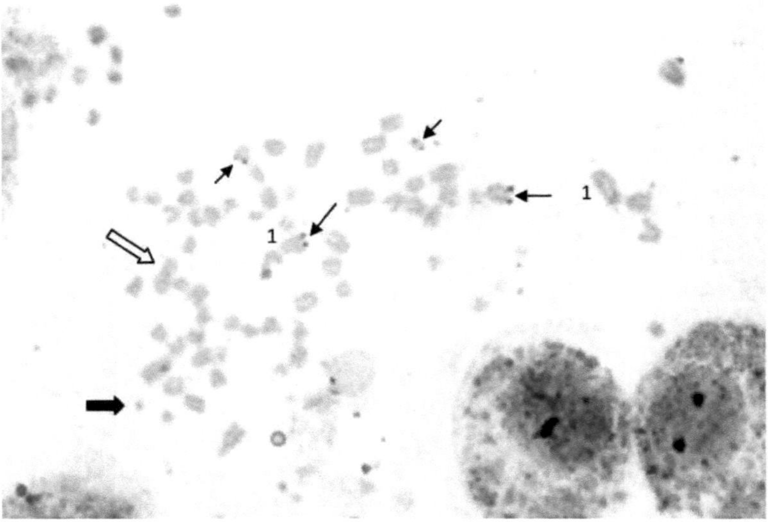

Figure 42 - Metaphase of tumor tissue. Male Boxer dog (2n=78). Observation of silver marking in the telomeric region of four chromosomes (single arrows). The wide white arrow indicates the X sex chromosome and the wide black arrow indicates the Y sex chromosome; the first pair is indicated with the number 1. Silver nitrate staining (Ron Band). Visualization under Zeiss microscope (1000x) and green filter

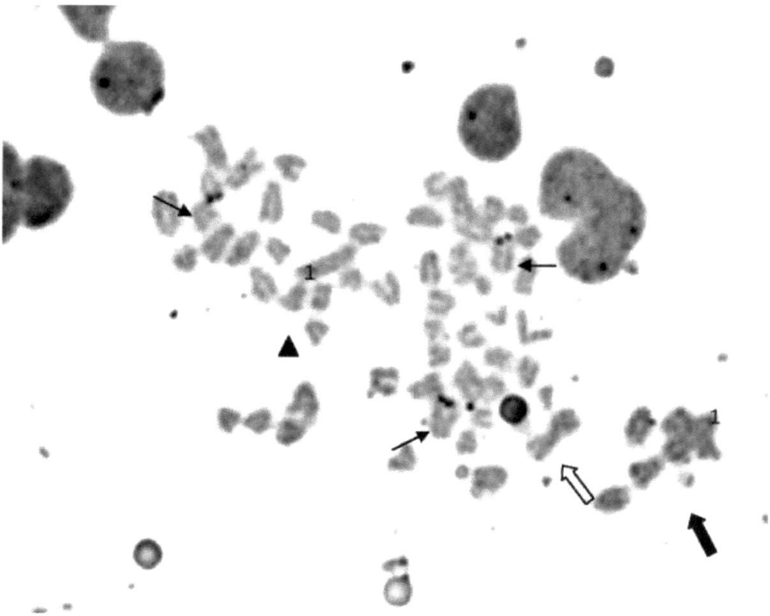

Figure 43 - Metaphase of tumor tissue. Male Boxer dog (2n=77). Observation of a submetacentric chromosome (triangle) and silver marking in the telomeric region of three chromosomes (single arrows). The wide white arrow indicates the X sex chromosome and the wide black arrow indicates the Y sex chromosome; the first pair is indicated with the number 1. Silver nitrate staining (Ron Band). Visualization under Zeiss microscope (1000x) and green filter

7 DISCUSSION

The results of a retrospective study on the incidence of mastocytomas in dogs between July 1999 and July 2003, carried out at the Pathology Department of HOVET - FMVZ - USP, showed that 2.3% of the reports (94) were related to cases of mastocytomas in dogs, of which 53 animals (56.4%) were male. Ages ranged from one to 15 years, with an average of 8.13 ± 3.02; 58.6% of the dogs were between seven and ten years old. These data corroborate the results of a study carried out by Vail (1996), who showed an incidence of mastocytomas mainly in dogs with an average age of 8-9 years.

In addition, the results of a retrospective study carried out at the Pathology Department of HOVET - FMVZ - USP, indicated a higher prevalence of mastocytomas in animals without a defined breed (37.2%), Boxers (27.7%) and Poodles (7.4%), which coincides with reports that the racial predisposition involves mostly dogs without a defined breed, but also affects the Boxer, Boston Terrier, English Bulldog, English Bull Terrier, Beagle, Labrador and Schanauzer breeds (VAIL 1996, O'KEEFE 2003).

According to Patnaik's (1984) classification, 16% of the tumors were grade I, 26.6% were grade II, 16% were grade III and 4.3% were grade II and III in the same animal. Patnaik (1984) presented the incidence of mastocytomas in terms of grade, with 36% classified as grade I, 43% as grade II and 20% as grade III.

A comparative study was also carried out to assess the number of chromosomes. Blood samples and tumor masses were analyzed from five dogs of the Boxer, Pug and Labrador breeds, as well as two dogs with no defined breed. In the analysis carried out, the chromosome count in the blood showed the usual number (2n=78), however, there was a variation between 74 and 79 chromosomes in the tumor. These results corroborate studies carried out on canine TVT, where the karyotype showed a numerical variation between 57 and 64 chromosomes (COHEN, 1978).

A survey of histopathological examinations carried out by the Animal Pathology Service at HOVET - FMVZ - USP between August 2004 and April 2008 revealed 206 cases of mastocytomas in dogs. With regard to breed, 31% of cases occurred in Boxer animals, followed by 24% of animals with no defined breed, 8% of Labradors and the remainder corresponding to the total of other breeds (37%), data which has also been confirmed by previous studies (VAIL 1996, O'KEEFE 2003).

The age of the dogs ranged from 3 to 15 years, with a prevalence of 42% of animals aged between 7 and 10 years, which coincides with data reported in the literature.

With regard to cytogenetics, numerical and structural alterations were found in the analyses carried out on the chromosomes in mitotic metaphase of the three Boxer breed animals.

Among the chromosomal aberrations, metacentric and submetacentric chromosomes were found, as well as HSR. The presence of metacentric and submetacentric chromosomes, in addition to sex chromosomes, indicates the occurrence of centric fusions of acrocentric chromosomes.

In the analysis of tumor cell metaphases, among all the stains, the first changes were observed. Of the 352 metaphases analyzed, 70.9% showed normal diploid numbers, 2n=78. Another 17.4% of the metaphases showed 2n=77; 9.8% had 2n=76. Five metaphases (1.3%) had 2n=79 and two cells had 2n=75.

Some studies on neoplasms have reported numerical and structural changes in canine cariocytes, such as a study on mastocytomas by Stone et al. (1991). The authors cytogenetically evaluated tumor cells, which showed cells with hyperdiploidy and hypodiploidy, an increase in the number of metacentric chromosomes, changes in configuration and cells with the loss of an X chromosome. This information authenticates the results of this dissertation.

Other studies have been carried out, such as in cases of mammary carcinomas, where analysis of metaphases using conventional staining revealed cases of chromosomal hypodiploidy, hyperdiploidy and tetraploidy (OWEN et al., 1977, 1981; ELSE et al., 1982; NORVALL et al. 1984; WOLFE et al., 1986).

Two cases of canine osteosarcoma were analyzed cytogenetically and revealed numerical and structural alterations with various centric fusions identified on chromosomes 1/3, 1/30 and 2/19, 3/19, 12/18, 13/17 (MAYR and ESCHBORN, 1991).

Mayr et al. (1990) cytogenetically characterized a canine hemangiopericytoma, which showed a reduced number of chromosomes (2n=74), in addition to structural alterations such as deletions and centric fusions involving chromosomes 5/6, 5/14, 7/15 and 9/17, as well as a deletion in chromosome 1.

The cytogenetic evaluation of breast tumors found trisomy 1 in 20% of the metaphases studied and monosomy X together with trisomy 1 in 30% of the cells in metaphase (MAYR et al., 1991).

In 1992, Mayr and Kramberger-Kaplan cytogenetically analyzed canine mammary sarcomas and found that 60% of the metaphases observed showed a normal karyotype, while the rest showed structural and numerical alterations.

In 1994, Mayr analyzed four solid tumors in dogs and found structural and numerical changes in the composition of the karyotype of these animals, with metacentric chromosomes, centric fusions and translocations.

Through cytogenetic analysis of canine osteosarcomas, Mayr et al. (1991) detected numerical and structural alterations in all the metaphases observed. There were 1/3, 1/30, 2/9 centric fusions, as well as chromosome counts of 65 to 67. All the cells showed centric fusions 3/19, 3/?, 12/18 and 13/17.

The cytogenetic analysis of a case of canine malignant melanoma showed that half of the metaphases analyzed had normal canine chromosomes (2n=78), with no alterations detected. However, the other metaphases observed had a count of 79 to 81 chromosomes. In these hyperdiploid cells, centric fusions were observed, partially identified as 6/?, 10/? and 11/? (MAYR et al., 1992).

Basic studies related to canine oncology are attractive due to the great demand for new diagnostic and therapeutic methods in veterinary medicine. In this sense, cytogenetic techniques have great potential, as they allow a broad understanding of tumor behavior and biology. Thus, it is possible to understand the mechanisms that cause genetic alterations during the carcinogenic process, and in the future, they will be useful as an important tool for assessing the prognosis of animals with neoplasms.

8 CONCLUSIONS

- No numerical changes were observed in metaphases from peripheral lymphocytes.
- There were numerical and structural changes observed in the metaphases of mastocytoma cells.
- Numerical and structural chromosomal alterations were detected in mitotic metaphases originating from tumor tissues.
- The C and G band staining techniques provided a better analysis of the metaphases.
- Structural alterations may be associated with the degree of malignancy of tumors as a result of an attempt to repair genetic errors caused by uncontrolled cell division.

REFERENCES

ARMITAGE, P.; DOLL, R. The age distribution of cancer and a multi-stage theory of carcinogenesis. **Britsh Journal of Cancer**, v.8, n.1, 1954.

CARROLL, S. M.; GAUDRAY, P.; DE ROSE, M. L.; EMERY, J. F.; MEINKOTH, J. L.; NAKKIM, E.; SUBLER, M.; VON HOFF, D. D.; Characterization of an episome produced in hamster cells that amplifies a transfected CAD gene at high frequency: functional evidence for a mammalian replication origin. **Molecular Cell Biology**. v 7, n. 5, p.1740-50, 1987.

CARROLL, S. M.; DE ROSE, M. L.; KOLMAN, J. L.; NONET, G. H.; KELLY, R. E.; WAHL, M. Localization of a bidirectional DNA replication origin in the native locus and in episomally amplified murine adenosine deaminase loci. **Molecular Cell Biology**. n. 13, v. 5, p. 2971-81, 1993.

CHIARELLI, B. The chromosome complement of Nasalis larvatus. Experientia. v. 22, n. 12, p.797, 1966.

COHEN, D. The transmissible venereal tumor of the dog- a naturally occurring allograft? A review. In: **WEISS, D. W.** Ed. **Immunological parameters of host-tumor relationships.** New York: Academic, v.5, p.14-19, 1978.

COTRAN, R. S.; KUMAR, V.; ROBBINS, S. L. Neoplasia. In: **Robbins pathologic basis of disease.** 4. ed. Philadelphia: W. B. Saunders, 1989. p. 239-305.

De VINNEY, R. GOLD, W.M. Establishment of two dog mastocytoma cell lines in continous culture. **American journal of respiratory cell and molecular biology**, v. 3, n. 3, p.413-420, 1990.

ELSE, R. W.; NORVAL, M.; NEILL, W. A. The characteristics of a canine mammary carcinoma cell line, REM 134. **Britsh Journal of Cancer**. V.46, v. 4, p.675-81, 1982.

FARBER, E. The multistep nature of cancer development. **Cancer Research**, Baltimore, v. 44, p. 4217-4223, 1984.

FARBER, E.; SARMA, D. S. R. Biology of disease. Hepatocarcinogenesis: a dynamic cellular perspective. **Laboratory Investigation**, New York, v. 56, n. 1, p. 4-22, 1987.

FORD, C. F.; HAMERTON, J. L. A colchicin hypotonic citrate squash sequence for mammalian chromosomes. **Stain Thechnology**. v. 5, n. 31, p. 247-251. 1956.

FUJINAGA, T.; YAMASHITA, M.; YOSHIDA, M. C.; MIZUNO, S.; OKAMOTO, Y.; TAJIMA, M.; OTOMO, K. Chromosome analysis of canine transmissible sarcoma cells. **Zentralblatt für Veterinarmedizin. Reihe C: Anatomie, Histologie, Embryologie**. v. 36, n. 7, p. 481-489, 1989.

GAUBATZ, J. W. Extrachromosomal circular DNAs and genomic sequence plasticity in eukaryotic cells. **Mutation Research**. v. 237n. 5-6, p. 271-292, 1990.

GOLD, W. M. Isolated canine mastocytoma cells: porpagtion and characterizations of two cell lines. **American Journal Physiology**. 1986.

GOLDSCHMIDT, M. H.; SHOFER, F.S. Skin tumors of the dog and cat. 1ª ed **Oxford**: Pergamon Press, 1992. 316p.

GRAHAN, J. C.; O'KEEFE, D. A. Soft Sarcomas and Mast cell tumors. In: Birchard, S. J., Scherding, R. G. **Saunders Manual of Small Animal Practice**, Philadelphia: W. B. Saunders, 1994, 200:206.

GRAHAN, J. C., O'KEEFE, D. A.: Soft tissue sarcomas and mastocytomas. In: BICHARDS, S.J.; SHERDING, R. G. **Manual saunders: Clinica de pequenos animais** 1 ed. Roca: Sâo Paulo, 1998. p.91-100.

GROSS, L. T.; IHRHE, P. J.; WALDER, E. J. **Veterinary dermatopathology: a macroscopic and microscopic evaluation of canine and feline skin disease**. St. Louis: Mosby, 1992. 519p.

Guerra, M. **Introduction to General Cytogenetics**. Ed. Guanabara Koogan: Rio de Janeiro-RJ. 142p. 1988).

GUERRA, M., SOUZA, M. J. **Como observar cromossomos: um guia de técnicas em cytogenética vegetal, animal e humana**. Ed. Guanabara Koogan: Rio de Janeiro-RJ p. 89, 2002.

GRAPHODATSKY, A. S.; BEKLEMISHEVA, V. R.; DOLF, G. High-resolution GTG- banding patterns of dog and silver fox chromosomes: description and comparative analysis. **Cytogenetics and cell genetics**. v. 69, n. 3-4, p. 226-231,1995.

HARE, P. J. Malignant haemangio-endothelioma. Proceedings of the Royal Society of Medicine. n. 58, p. 423, 1965.

HOLLIDAY, R. Chromosome error propagation and cancer. **Trends in genetics**. n. 5, p.4245, 1989.

Howard-Peebles PN, Pryor JC. The R-banding pattern of the canine karyotype. **The Journal of heredity**. v. 71, n. 5, p. 361-2, 1980.

HOWELL, W. M.; BLACK, D. A. Controlled silver-staining of nucleolus organizer regions with protective colloidal developer: a 1st step method. **Experientia.** n. 36, p. 1014-1015, 1980.

HSU, T. C.; MOORHEAD, P. S. Chromosome anomalies in human neoplasms with special reference to the mechanisms of polyploidization and aneuploidization in the HeLa strain.

Annals of the New York Academy of Sciences. v.63, n. 6, p. 1083-1094, 1956.

HUEPER, W.C. Enviromental carcinogenesis in man and animals. **Annals of the New York Academy of Sciences.** v. 108 p. 963-970, 1963.

ISCN, An International System for human cytogenetic nomenclature (Mitelman F, eds). **Basel: S. Karrer**. 1995

JANEWAY JR, C. A.; TRAVERS, P.; WALPORT, M.; SHLOMCHIK, M. J. Humoral immune response. In: **Immunobiology:** the immune system in health and disease. 5. ed Sâo Paulo: Artmed. 2002. p. 399-400.

LANGFORD, C. F.; FISCHER, P. E.; BINNS, M. M.; HOLMES, N. G.; CARTER, N. P.; Chromosome-specific paints from a high high-resolution flow karyotipe of the dog. **Chromosome Research**. n. 4, p. 115-123, 1996.

LAZARUS, S. C. Role of inflammation and inflammatory mediators in airways disease. **The American Journal of Medicine** v. 81, n. 5ª, p. 2-7, 1986.

LOWER, K. Malignant and Benign Neoplasms. In: MEDLEAU, L.; HNILICA, K. A.

Small Animal Dermatology: Color Atlas and Therapeutic Guide. 1.ed. Sâo Paulo: Roca, 2003. 353 p.

KITCHELL, B. E. Cancer therapy for geriatric dogs and cats. **Journal of the American Animal Hospital Association**, v.29, p.41-8, 1993.

KRAEGEL S. A., PAZZI KA, MADEWELL BR. Sequence analysis of canine p53 in the region of exons 3-8. **Cancer letters**. v.92, n. 2, p. 181-186, 1995.

LEMARIÉ, R. J., LEMARIÉ, S. L., HEDLUND, C. S. Mast cell tumors: clinical management. **Compendium on Continuing Education for the practicing Veterinarian**, v9, n.9, p. 10851101, 1995.

LOWER, K. Malignant and Benign Neoplasms. In: MEDLEAU, L.; HNILICA,

K. A. **Small Animal Dermatology**: Color Atlas and Therapeutic Guide.

1.ed. Sâo Paulo: Roca, 2003. 353 p.

MACY, D.W.; MACEWEN, E.G. Mast cell tumors. In: WITHROW S.J.; MACEWEN, E.G (Ed.). **Veterinary oncology**. Philadelphia: J.B. Lippincott, 1986. p.156-165.

MADEWELL, B.R.; THEILEN, G. H.: Mast cell and melanocytic neoplasms. In: THEILEN, G.

H.; MADEWELL, B. R.: **Veterinary Cancer Medicine**, 2nd edition, Philadelphia: Lea & Febiger, 1987, 310-4.

Banding analysis of the somatic chromosomes of the domestic dog (Canis familiaris). Manolache M, Ross WM, Schmid M. **Canadian journal of genetics and cytology.**

Journal canadien de génétique et de cytologie. v. 18, n. 3, p. 513-8, 1976.

MAYR, B. Cytogenetic analysis of four solid tumors in dogs. **Research in Veterinary Science**. 57, 88-95. 1994.

MAYR, B.; ESCHBORN, U. Characterization of complex karyotype in two canine bone tumours. **Research in Veterinary Science.** n. 51, p.341-3, 1991.

MAYR, B.; GILLI, H.; SCHLEGER, W.; REIFINGER, M; BURTSCHER, H. Cytogenetic characterization of mammary tumours in two domestic dogs. **Journal of veterinary medical education**. n. 38, p. 141-7, 1991.

MAYR, B.; KRAMBERGER-KAPLAN, E. Analysis of complex alterations in three canine mammary sarcomas. **Research in Veterinary Science.** n. 53, p. 205-11, 1992.

MAYR, B., REIFINGER, M., BREM, G., FEIL, C., SCHLEGER, W. Cytogenetic, ras, and p53: srudies in cases of canine neoplasms (hemangiopericytoma, mastocytoma, histiocytoma, chloroma). **The American Genetic Association.** n. 90, p. 124-128, 1999.

MAYR, B.; SWIDERSKY, W.; SCHLEGER, W.; REIFINGER, M. Cytogenetic characterization of canine haemangiopericytoma. **The British veterinary journal.** n.146, p. 260, 1990.

METCALFE, D. D., BARAM, D., MEKORI, Y. A. Mast cells. **Physiological reviews.** v. 77, n. 4, p. 1033-1079, 1997

MILLER, D. M. The occurrence of mast cell tumors in young Shar- Peis.

Journal of veterinary diagnostic investigation. v.7, p.360- 363, 1995.

MISDORP, W. Mast cells and canine mast cell tumors. A review. **The Veterinary quarterly.** v. 26, n. 4, p. 156-69, 2004.

MITELMAN, F.; JOHANSSON, B.; MERTENS, F. (eds): Mitelman **Database of chromosome aberrations in cancer**.

(http://cgap.nci.nih.gov/Chromosomes/Mitelman) (2007).

MOORE, W. JR.; LAMBERT, P. D. The chromosomes of the beagle dog.

The Journal of heredity. n. 54, p.273-276,1963.

MURPHY, G. F.; SELLHEYER, K.; MIHM Jr, M. C. Diseases of the organ systems. In:

KUMAR, V.; ABBAS, A. K.; FAUSTO, N. **Pathologia** - bases patológicas das doenças. 7. ed Rio de Janeiro: Elsevier. 2005. p.1306-1308.

WOLFE, L. G.; SMITH, B. B.; TOIVIO-KINNUCAN, M. A.; SARTIN, E. A.; KWAPIEN, R. P.; HENDERSON, R. A.; BARNES, S. Biologic properties of cell lines derived from canine mammary carcinomas. **Journal of the National Cancer Institute**. v. 77, n. 3, p. 783-792,

1986.

O'BRIEN, M. G.; WHITHRON, S. J. Surgical oncology. In: ETTINGER, S. J.; FELDMAN, E.C. **Treatise on veterinary internal medicine**. 4. ed Sâo Paulo: Manole, 1997. p. 723-32.

OGILVIE, G.; MICHENER,K. Care beyond cure: mast cell tumor-New tricks for the new millennium. In: **Midwest veterinary conference**, 116, Columbus, 2000. Proceedings. Columbus, 2000. v2, p. 55-8.

OGILVIE, G. K.; MOORE, A. S. Mast cell tumors in Dogs. In:. Managing the Veterinary Cancer Patient: a practice manual. 1.ed. **Trenton**: Vet. Learning Systems Company, 1995. p.493 -510.

O'KEEFE, D. A. Soft tissue sarcomas and mastocytomas In: **Clinica de pequenos animais, manual Saunders,** 2003 p. 241 - 4.

OWEN, L. N,; MORGAN, D. R,; BOSTOCK, D. E,; FLEMANS, R. J. Tissue culture and transplantation studies on canine mammary carcinoma. **European journal of cancer**. v. 13, n. 2, p. 1445-9, 1977.

PAINTER, T. S. CHROMOSOME NUMBERS IN MAMMALS. **Science**. v. 17, n. 61, p. 423-424, 1925.

PATNAIK, A. K.; EHLER, W.J.; MACEWEN, E. G.: Canine cutaneous mast cell tumor: morphologic grading and survival time in 83 dogs, **Veterinary Pathology**, v. 21, n. 5; p. 469-474, 1984.

PITOT, H. C.; DRAGAN, Y. P. Facts and theories concerning the mechanisms of carcinogenesis. **The FASEB journal**. v. 5, n. 9, p. 2280-2286, 1991.

POULSEN, B. S.; SHIBASAKI, Y.; IKEUCHI, T.; R0NNE, M. Banding studies in Canis familiaris. Replication patterns in karyotypes from lymphocyte cultures. **Cytobios.** v. 62, n. 250-251, p. 161-165, 1990.

PULLEY, L.T.; STANNARD, A.A. Tumors of the skin and soft tissues. In: MOULTON, J.E. **Tumours In Domestic Animals**. Berkeley: California Press, 1990. p.23-87.

REDDY, K. S. Double minutes (dmin) and homogeneously staining regions (hsr) in myeloid disorders: a new case suggesting that dmin forms hsr in vivo. **Cytogenetic Genome Research**. n. 119, p. 53-59, 2007.

N. REIMANN, S. BARTNITZKE, I. NOLTE, AND J.BULLERDIEK. Working with canine chromosomes: current recommendations for karyotype description. **The American Genetic Association**. n. 90, p. 31-34, 1999.

REGUERA, M. J., RABANAL, R. M., PUIGDEMONT, A., FERRER, L. Canine mast cell tumors express stem cell factor receptor. **The American Journal of dermatopathology**. v. 22, n. 1, p. 49-54, 2000.

SEABRIGHT, M. A rapid banding technique for human chromosomes. **Lancet. 2**: 971, 1971.

SCOTT, *W.* D.; MILLER, H. *W.;* GRIFFIN, E. C. Neoplastic and non-neoplastic tumors. In: MILLER and KIRK. **Small Animal Dermatology**. 5. ed. Rio de Janeiro: Interlivros, 1996. Translation of: Small Animal Dermatology.

SELDEN, J. R.; MOORHEAD, P. S.; OEHLERT, M. L.; PATTERSON, D. F. The Giemsa banding pattern of the canine karyotype. **Cytogenetics and cell genetics**. v.15, n. 6, p. 380-388, 1975.

SELYE,H.: Mast Cell - **Review**, 1st Edition, Washington: Butterworths Inc., 1965, p. 275-78.

SIMOES, J. P. C. Canine mast cell tumors: a comparison of staining techniques. **Journal of Veterinary Diagnostic Investigation**, v. 06. p. 458 - 65, 1994.

SIMPSON, A. M.; LUDWIG, L. L.; NEWMAN, S. J. et al. Evaluation of surgical margins required for complete excision of cutaneous mast cell tumors in

Dogs. **Journal of the American Veterinary Medical Association**. n. 224, p. 236 - 240, 2004.

SOLOMON, E.; BORROW, J.; GODDARD, A. D. Chromosome aberrations and cancer. **Cancer Genetic and Cytogenetic**. v. 8, n. 5035, p. 1153, 1991.

SOTER, N. A. Mastocytosis and the skin. **Hematology/Oncology Clinics of Noth America**, v. 14, n. 3, p.537-555, 2000.

SOUZA, T. M.; FIGUERA, R. A.; IRIGOYEN, L. F.; BARROS, C. S. Retrospective study of 761 skin tumors in dogs. **Ciência Rural**. v.36, n.2, p. 555 - 560, 2006.

STANNARD, A. A.; PULLEY, L. T. Tumors of the skin and soft tissues. In: MOULTON, J. E. **Tumors in domestic Animals**. 3. ed. Berkeley: University of California, 1990. chap. 2, p. 23 - 87.

STONE, D.; JACKY, P. B.; PRIEUR, D. J.. Cytogenetic evaluation of four canine mast cell tumors. **Cancer Genetic and Cytogenetic.** n. 53, p. 105-112, 1991.

SUMNER, A. T. A simple technique for demonstrating centromeric heterochromatin . **Experimental cell research**. n. 75, p. 304-306, 1972.

SWITONSKI, M.; REIMANN, N.; BOSMA, A. A.; LON, S.; BARTNITZKE, S.;

PIENKOWSKA, A.; MORENO-MILAN, M. M.; FISCHER, P. Report on the progress of the standardization of the G-banded canine karyotipe. **Chromosome Research**. n. 4, p. 306-309, 1996.

VAIL, D. M.: Mast cell tumors. In: WITHROW S. J.; MACEWEN, E. G.: **Small Animal Clinical Oncology**, 2nd ed. Philadelphia PA: WB Saunders, 1996. p. 192:210.

WINKLER, S., MURUA ESCOBAR, H., REIMANN-BERG, N., BULLERDIEK, J., NOLTE, I. Cytogenetic investigations in four canine lymphomas. **Anticancer Reserach**. v. 25, n. 6B, p. 3995-3998, 2005.

WHITE, M. J. D. Os Cromossomos. Companhia **Editora Nacional**, Sao Paulo - SP. 196p. 1977

ZAPPULLA, J. P., AROCK, M., MARS, L. T., LIBLAU, R. S. Mast cells: new targets for multiple sclerosis therapy? **Journal of neuroimmunology**. v.131, n. 1-2, p. 1520, 2002.

ZEMKE, D.; YAMINI, B.; YUZBASIYAN-GURKAN, V. Mutations in the Juxtamembrane Domain of c -KIT are Associated with Higher Grade Mast Cell Tumors in Dogs. **Vetherinary. Pathology**. v.39, p. 529 -535, 2002.

WAHL, G. M. The importance of circular DNA in mammalian gene amplification. Cancer Research. v. 49, n. 6, p. 1333-1340, 1989.

WANG HC, FEDOROFF S. Banding in human chromosomes treated with trypsin. **Nature: New biology.** v. 235, n. 54, p. 52-54, 1972.

MIX
Papier aus verantwortungsvollen Quellen
Paper from responsible sources
FSC® C105338

Printed by Books on Demand GmbH, Norderstedt / Germany